Springer Theses

Recognizing Outstanding Ph.D. Research

Aims and Scope

The series "Springer Theses" brings together a selection of the very best Ph.D. theses from around the world and across the physical sciences. Nominated and endorsed by two recognized specialists, each published volume has been selected for its scientific excellence and the high impact of its contents for the pertinent field of research. For greater accessibility to non-specialists, the published versions include an extended introduction, as well as a foreword by the student's supervisor explaining the special relevance of the work for the field. As a whole, the series will provide a valuable resource both for newcomers to the research fields described, and for other scientists seeking detailed background information on special questions. Finally, it provides an accredited documentation of the valuable contributions made by today's younger generation of scientists.

Theses are accepted into the series by invited nomination only and must fulfill all of the following criteria

- They must be written in good English.
- The topic should fall within the confines of Chemistry, Physics, Earth Sciences, Engineering and related interdisciplinary fields such as Materials, Nanoscience, Chemical Engineering, Complex Systems and Biophysics.
- The work reported in the thesis must represent a significant scientific advance.
- If the thesis includes previously published material, permission to reproduce this must be gained from the respective copyright holder.
- They must have been examined and passed during the 12 months prior to nomination.
- Each thesis should include a foreword by the supervisor outlining the significance of its content.
- The theses should have a clearly defined structure including an introduction accessible to scientists not expert that particular field.

More information about this series at http://www.springer.com/series/8790

Karin J. Bichler

Morphology and Dynamics of Bottlebrush Polymers

Doctoral Thesis accepted by Louisiana State University, USA

Springer

Karin J. Bichler
Department of Chemistry
Louisiana State University
Baton Rouge, LA, USA

ISSN 2190-5053 ISSN 2190-5061 (electronic)
Springer Theses
ISBN 978-3-030-83378-7 ISBN 978-3-030-83379-4 (eBook)
https://doi.org/10.1007/978-3-030-83379-4

This Springer imprint is published by the registered company Springer Nature Switzerland AG
The registered company address is: Gewerbestrasse 11, 6330 Cham, Switzerland

Supervisor's Foreword

Showing properties entirely different from those of linear polymers, bottlebrush polymers are highly appealing for fundamental research. Owing the name their specific structure—branch-like side chains connected to a linear backbone—these macromolecules appear to be molecular replica of bottlebrushes used in households. Materials made out of bottlebrush polymers exhibit extraordinary features, including super softness, hyperelasticity, or ultralow viscosities, beyond those properties usually achievable by additives such as plasticizers or fillers to linear polymers. Dr. Karin Bichler's dissertation is among the first studies of these macromolecules targeting the understanding of their macroscopic properties from the molecular perspective.

Ultimately molecular interactions determine both the materials' behavior and structure and dynamics at the nanoscale. Hence, exploring bottlebrush polymers at the molecular level is the key of directed materials' design, unleashing the full potential of these materials. Dr. Karin Bichler's dissertation provides deeper insight in the morphology and dynamics of bottlebrush polymers in different states. More specifically, the main objective of the dissertation is the understanding of ultrasoft branched poly(dimethylsiloxane) polymers in different environments, including polymer melts and solutions.

At the high grafting densities studied by Dr. Karin Bichler, bottlebrush polymers are observed. Unlike the random coil conformation of polymers of linear architecture, bottlebrushes can be spherical or elongated objects. Using a backbone based on one of the most flexible polymers, poly(dimethylsiloxane) (PDMS), this is already a key finding, because it illustrates the emergence of the bottlebrush morphology as a consequence of the stretched backbone caused by the interactions of densely packed side chains, ultimately leading to the fascinating properties of bottlebrushes. The importance of this fact lies in the freedom of choice of the backbone, which could enable independent tuning of properties related to chemical and physical parameters, such as solubility.

Novel opportunities for directed materials design have been generated, because bottlebrushes seem to retain their conformation in different environments, including solvents and polymer melts of linear architecture. The miscibility of shell-only particles and environment can be tuned via molecular interactions. This enables a better control over the spatial arrangement of the shell-only particles, ranging from colloidal structures to agglomerates at higher concentrations. By identifying these structural parameters Dr. Karin Bichler continued to unleash the potential of bottlebrushes in directed materials design.

Material properties strongly depend on molecular dynamics. This is particular important for bottlebrush polymers, because the relaxation at the length scale of segments strongly depends on the length of the side chains. Dr. Karin Bichler found a change of up to 3 orders in time scale that accompanies the gradual transition from highly confined segmental dynamics to a behavior that indicates a transition to unconfined motion, from low to high molecular weights, respectively. At the length scale of the polymer chain, the signature of unentangled polymer melt dynamics has been identified in higher molecular weight bottlebrush.

Understanding molecular dynamics is the key to accomplish a better understanding of macroscopic properties to ultimately control the materials behavior. The dissertation of Dr. Karin Bichler presents the first study that takes advantage of the spatial resolution of neutron spectroscopy to elevate fundamental understanding of bottlebrush polymers and beyond. The obtained mean square displacement conveniently distinguishes between several processes, including segmental relaxation, methyl group rotation, and fast dynamics typically associated with the Debye-Waller factor. The combination of multiple instruments enabled the isolation of different processes in the time and spatial window and finally permitted to use time-temperature scaling to generate a mean square displacement over 7 orders of magnitude in time. This procedure is a more universal version of the time-temperature superposition principle, delineated at the molecular length scale and valid for polymers of increasingly complex architectures, despite their structural and dynamical heterogeneities, and even below the glass transition temperature. This new procedure is highly transformative and can be used to apply this generalized time-temperature superposition procedure to a variety of characterization techniques and materials.

The dissertation of Dr. Karin Bichler is among the first holistic studies of bottlebrush polymers, involving morphology and dynamics at the length scale of the polymer chain and below. The key finding of the existence of bottlebrushes based on a flexible polymer as backbone was brought to the next level by identifying significant changes of the dynamics. The results seem to be independent of the material, because they are essentially determined by individual parameters, like side chain density and length. These facts reflect the important contributions made by Dr. Karin Bichler toward the understanding of properties based on molecular parameters, which will enable directed materials design.

Department of Chemistry
Department of Physics and Astronomy
Louisiana State University
Baton Rouge, LA, USA
June 2021

Gerald J. Schneider

Parts of This Dissertation Have Been Published in the Following Journal Articles

1. K.J. Bichler, B. Jakobi, V. García Sakai, A. Klapproth, R.A. Mole, G.J. Schneider, Universality of the time-temperature scaling observed by neutron spectroscopy on bottlebrush polymers. Nano Lett. **21**(10), 4494–4499 (2021). https://doi.org/10.1021/acs.nanolett.1c01379
2. K.J. Bichler, B. Jakobi, G.J. Schneider, Dynamical comparison of different polymer architectures—bottlebrush vs. linear polymer. Macromolecules **54**(4), 1829–1837 (2021). https://doi.org/10.1021/acs.macromol.0c02104
3. K.J. Bichler, B. Jakobi, V. García Sakai, A. Klapproth, R.A. Mole, G.J. Schneider, Short time dynamics of PDMS-g-PDMS bottlebrush polymer melts investigated by quasi-elastic neutron scattering. Macromolecules **53**(21), 9553–9562 (2020). https://doi.org/10.1021/acs.macromol.0c01846
4. B. Jakobi, K.J. Bichler, A. Sokolova, G.J. Schneider, Dynamics of PDMS-g-PDMS bottlebrush polymers by broadband dielectric spectroscopy. Macromolecules **53**(19), 8450–8458 (2020). https://doi.org/10.1021/acs.macromol.0c01277
5. K.J. Bichler, B. Jakobi, S.O. Huber, E.P. Gilbert, G.J. Schneider, Structural analysis of ultrasoft PDMS-g-PDMS shell-only particles. Macromolecules **53** (1), 78–89 (2020). https://doi.org/10.1021/acs.macromol.9b01598

Acknowledgments

This work would not have been possible without the help and support of many people. I am grateful that I had the opportunity to do my dissertation at Louisiana State University and, therefore, would like to thank all the people who contributed to the work.

First, I would like to thank my supervisor Prof. Dr. Gerald Schneider for encouraging me to join LSU and to start the research of my dissertation in his group. Also, I am thankful for his help and support during all these times and beyond as well as for the nice and friendly atmosphere in the lab and the office.

Also, I would like to thank Prof. Dr. Phillip Sprunger, Prof. Dr. John DiTusa, and Prof. Dr. Shyam Menon for being part of my PhD committee and their interest and help in my research.

All the neutron experiments at ANSTO (Australia), MLZ (Germany), and ISIS (United Kingdom) would not have been successful without the instrument scientists. Therefore, I would like to thank Dr. Alice Klapproth, Dr. Anna Sokolova, Prof. Dr. Elliot Gilbert, and Dr. Richard Mole at ANSTO, Dr. Aurel Radulescu at FRMII, and Dr. Victoria García Sakai and Dr. Ian Silverwood at ISIS for their help and support during and after the experiments.

Furthermore, I would like to thank Bruno Jakobi and Stefan Huber for the good cooperation and all the helpful advice, as well as the good friendship which has been developed over the years. Also, I would like to thank all members of the Schneider group.

In addition, I would like to thank the Physics & Astronomy and the Chemistry Department of Louisiana State University, especially Prof. Dr. Robert Hynes and Paige Whittington, for all the helpful answers and advice for all kind of questions.

For the major financial support for this dissertation, I would like to thank the US Department of Energy (DOE), grant DE-SC0019050, and for the first neutron scattering experiment, I would like to thank the Louisiana Consortium for Neutron Scattering (LaCNS) under EPSCoR grant DE-SC0012432 with the additional support from Louisiana Board of Regents.

Finally, I would like to thank my parents, my sister, and the rest of my family for their support and understanding during my dissertation and beyond.

Contents

Chapter 1
Introduction

Abstract Bottlebrushes belong to the family of branched polymers with a high grafting density of side chains, covalently bonded to the backbone. Due to steric repulsions, elongated or spherical shapes occur instead of random coil conformations, known from linear polymers. Besides these morphological changes, bottlebrushes show a reduced entanglement density compared to their linear counterparts. Consequently, the dynamics change and super soft and hyperelastic materials result. These characteristics make bottlebrushes one important class of polymers connected with applications in several disciplines. Since the macroscopic behavior responsible for these applications is connected with the microscopic properties, it is important to understand the structure and the dynamical behavior of bottlebrush polymers.

Bottlebrush polymers belong to the family of branched polymers and have a high grafting density, i.e., high number of side chains [1–3]. Hereby, the point of attachment is called branch point [3, 4]. More in general, the bottlebrush regime is reached if the density of the grafted side chains reduces the ability of the backbone to adopt a random coil conformation, otherwise, a comb polymer is obtained (Fig. 1.1) [5].

Side chains and backbone can be synthesized separately and characterized in advance before the grafting reaction, leading to a well-customizable polymer. The material properties like stiffness and viscosity can be controlled over a much broader range than those achievable with other polymer architectures. These characteristics differ from those of linear polymers and allow a variety of applications in surface modification, material science or medicine [1, 2, 6–12]. Typical morphologies of bottlebrush polymers are cylindrical, or wormlike structures, caused by steric repulsion of the grafted side chains forcing the backbone to elongate [11, 13]. The extension is controlled by the length of the side chains and the grafting density [2, 14]. Bottlebrush polymers in their own melt can be described as thick filaments, whereby the whole system shows extraordinary rheological properties [15, 16]. In contrast to linear polymers, this resulted in unprecedented properties, like hyperelasticity or super softness [1].

The shape of bottlebrush polymers can be controlled by the ratio of the backbone to side chain molecular weight. If the backbone length decreases while the length of

K.J. Bichler, *Morphology and Dynamics of Bottlebrush Polymers*, Springer Theses,
https://doi.org/10.1007/978-3-030-83379-4_1

Fig. 1.1 Illustration of a comb polymer (left) and a bottlebrush polymer (right)

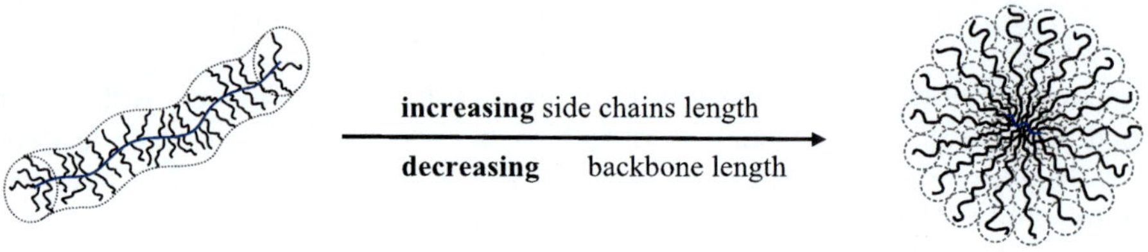

Fig. 1.2 Transition from cylindrical into spherical shape by changing the size ratio of backbone compared to side chains

the side chains increases, a transition from cylindrical into spherical objects occurs (Fig. 1.2) [8, 17, 18].

In the spherical region, the bottlebrush polymer is reminiscent of a star polymer with negligible core size, because all side chains are virtually focused to one point [19–21]. Furthermore, it can be compared with polymer grafted nanoparticles used as filler materials in polymer systems, whereby the side chains represent the grafted polymer, i.e., the shell of the particle. Therefore, a spherical bottlebrush polymer can be seen as a shell-only particle and as a model system in which the shell can be studied separately [22].

While the shape is influenced by the grafted side chains, also the dynamics of the chains are impacted. Up to now, only little is known about the dynamics of bottlebrush polymers. Most of the current knowledge comes from rheology, with a few supplements from Nuclear Magnetic Resonance (NMR) and dielectric spectroscopy experiments [3, 8, 15, 23–25].

For polymers with linear architecture or very short side chains, the relaxation is determined by the backbone [26]. If side chains exceed a certain length, rheology can detect a visible contribution of the side chains and two processes, based on the side chains and the backbone need to be distinguished [8].

For high molecular weight side chains, a rubbery plateau can be reached at short relaxation times [15]. This assumes that only the side chains entangle, while the bottlebrush itself is unentangled up to very high molecular weights [9, 15]. Furthermore, the motion of the branch points are slow and retard the relaxation of the backbone segments until all side chains can relax. Therefore, the side chains increase the longest relaxation time in the system due to high frictional effects. In contrast, the chain ends of the side chains, the dangling ends, relax faster compared to other segments. With an increasing number of dangling ends, the longest relaxation time decreases. So, depending on the side chain length and number of side chains, the longest relaxation time can either be longer or shorter compared to the linear counterparts [3]. Dielectric spectroscopy measurements on hetero bottlebrush polymers compared to their respective linear equivalents, revealed a slowed-down relaxation behavior of the side chains, possibly caused by the restrictions arising from the backbone and a faster relaxation of the backbone due to plasticizer effects [27]. The spin–spin relaxation time, T_2, for bottlebrushes, measured by NMR, changes as well, compared to the linear counterparts [28]. The backbone relaxation decreases with increasing side chain length and grafting density due to steric hindrances. For low molecular weight side chains, T_2 increases with increasing length. At a certain molecular weight, a decrease of T_2 was noted. These changes are explained by an increasing amplitude of the oscillation with increasing molecular weight of the side chains, resulting in a longer T_2, while the abrupt shortening of T_2 is explained by the interactions between longer side chains caused by excluded volume effects [28].

The viscoelastic and mechanical behavior of polymers is determined by their microscopic properties, i.e., the morphology and the dynamics [29]. As illustrated in Fig. 1.3, by means of the magnitude of the dynamic modulus, E, in relation to temperature (top) and time (bottom) in a polymer melt, it is obvious, that the dynamics extend over several orders of magnitude in time, from femtoseconds up to microseconds. This is a consequence of different molecular motions, like vibrations, segmental dynamics, chain dynamics, and diffusion, all happening on different length scales. At low temperatures, the polymer is in the glassy state and only vibrational motions or secondary relaxations are possible. By passing the glass transition temperature, the material starts to flow, and the segmental dynamics set in.

Increasing the temperature further results in the polymer dynamics, seen as the rubbery plateau for sufficiently high molecular weights. This is followed by center of mass diffusion in the long time limit [29].

One major property of bottlebrush polymers, for example, is the super softness, commonly attributed to long time scales. However, to understand these extraordinary properties within the bottlebrush polymers, the origin, i.e., the microscopic properties over all time scales need to be understood first.

Unlike all known literature, this dissertation concentrates on homo bottlebrush polymers based on poly(dimethylsiloxane) (PDMS), i.e., PDMS-g-PDMS bottlebrush polymers with a grafting density well inside the bottlebrush regime (Fig. 1.4). Here the side chains and the backbone have the same polymeric repeating unit as the linear counterparts. One advantage of using homopolymers is the reduced probability of unwanted phase separation or the formation of nano- or microdomains.

PDMS consists of inorganic Silicon-Oxygen (Si-O) bonds with a low rotational barrier resulting in a high flexibility, high mobility of the segments, and the entire molecule [30]. This makes PDMS very interesting for dynamical studies [30].

Furthermore, PDMS has one of the lowest glass transition temperatures, T_g, low viscosity, high gas permeability, high thermal stability, and low toxicity [31]. PDMS can be synthesized by anionic polymerization. This allows a customized chemical structure, including a molecular weight with a narrow molecular weight distribution,

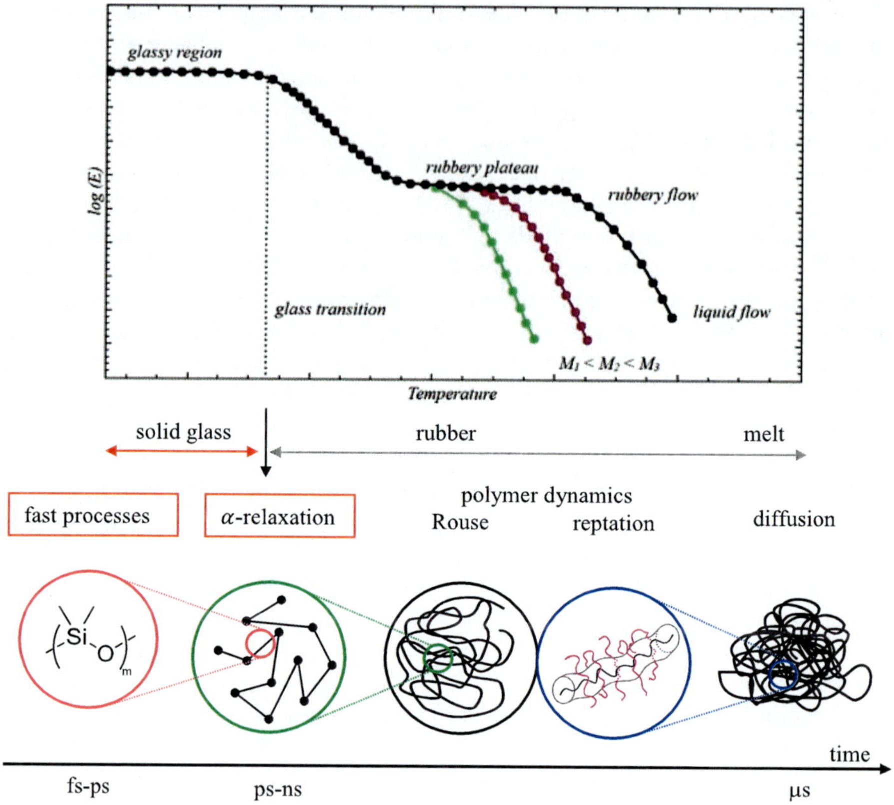

Fig. 1.3 Illustration of the contributions of different processes as they occur in a polymer melt. The different dynamical phenomena are connected to different molecular origins as displayed in the circles. This dissertation focuses on the short-time dynamics, i.e., fast processes and α-relaxation, as illustrated by the red boxes. (Adapted from Richter et al. [29])

Fig. 1.4 (**a**) Structure of poly(dimethylsiloxane) (PDMS) and (**b**) the resulting PDMS-g-PDMS bottlebrush polymer structure

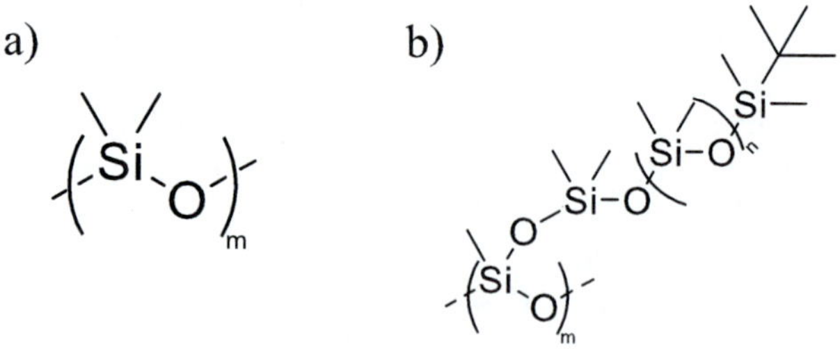

giving a well-defined polymer. This is especially important for neutron scattering experiments to simplify the analysis by minimizing the number of assumptions needed for the data description.

This dissertation concentrates on the morphology and the short-time dynamics, i.e., vibrations, methyl group dynamics, and segmental dynamics, of PDMS-g-

PDMS bottlebrush polymers, which are the foundations for further experiments exploring, e.g., the polymer dynamics in order to understand the physical properties.

Structural characteristics of bottlebrush polymers are in the length scale of angstrom to nanometer. Therefore, Small-Angle Neutron Scattering (SANS) is a suitable tool. The morphology of PDMS-g-PDMS bottlebrush polymers has been characterized in different environments, i.e., bottlebrush in good solvent, in good solvent with added linear chains and in linear host matrices, to follow the inherent structural changes. Since the shape of bottlebrushes depends on the size ratio of side chain to backbone length, samples with similar backbone length, but different side chains lengths have been studied to clarify if the known sphere to cylinder transition for relatively stiff bottlebrush polymers also occurs in homopolymers, i.e., PDMS-g-PDMS bottlebrush polymers.

The dynamical processes investigated here, vibrations, methyl group dynamics, and segmental dynamics, are in the time scale of ps to ns, which are well within the experimental time range of the used techniques, i.e., Broadband Dielectric Spectroscopy (BDS), Fast Field Cycling (FFC) Relaxometry, and Quasi-Elastic Neutron Scattering (QENS). The influence on the segmental relaxation times caused by attaching linear side chains to the backbone have been investigated by dielectric spectroscopy. In order to get insights, if the polymer dynamics are influenced by different architectures, fast field cycling measurements have been performed on samples, having similar molecular weight but different conformations. Since the morphology depends on the size ratio from backbone to side chains, the influences on the dynamics on these samples have been studied. Herefore, dielectric spectroscopy and QENS have been used. The latter technique results in a high energy resolution together with spatial resolution and allows to connect the dynamical behavior with geometrical constraints. In order to have a broad time window, which is necessary for polymers, a combination of three different neutron scattering instruments, neutron backscattering, and time-of-flight instruments, has been used. This allows to capture dynamics from vibrations up to segmental relaxations.

All polymers used in this dissertation were provided by Bruno Jakobi. As a part of his PhD work, the synthesis of the bottlebrush polymers, based on anionic ring opening polymerization, is described in his dissertation [32].

References

1. W.F. Daniel, J. Burdynska, M. Vatankhah-Varnoosfaderani, K. Matyjaszewski, J. Paturej, M. Rubinstein, A.V. Dobrynin, S.S. Sheiko, Solvent-free, supersoft and superelastic bottlebrush melts and networks. Nat. Mater. **15**(2), 183–189 (2016)
2. S.S. Sheiko, B.S. Sumerlin, K. Matyjaszewski, Cylindrical molecular brushes: synthesis, characterization, and properties. Prog. Polym. Sci. **33**(7), 759–785 (2008)
3. M. Abbasi, L. Faust, M. Wilhelm, Comb and bottlebrush polymers with superior rheological and mechanical properties. Adv. Mater. **31**(26), 1806484 (2019)

4. H. Ma, Q. Wang, W. Sang, L. Han, P. Liu, J. Chen, Y. Li, Y. Wang, Synthesis of bottlebrush polystyrenes with uniform, alternating, and gradient distributions of brushes via living anionic polymerization and hydrosilylation. Macromol. Rapid Commun. **36**(8), 726–732 (2015)
5. G. Polymeropoulos, G. Zapsas, K. Ntetsikas, P. Bilalis, Y. Gnanou, N. Hadjichristidis, 50th anniversary perspective: polymers with complex architectures. Macromolecules **50**(4), 1253–1290 (2017)
6. J. Paturej, S.S. Sheiko, S. Panyukov, M. Rubinstein, Molecular structure of bottlebrush polymers in melts. Sci. Adv. **2**(11), e1601478 (2016)
7. H. Liang, S.S. Sheiko, A.V. Dobrynin, Supersoft and hyperelastic polymer networks with brushlike strands. Macromolecules **51**(2), 638–645 (2018)
8. C.R. López-Barrón, P. Brant, A.P. Eberle, D.J. Crowther, Linear rheology and structure of molecular bottlebrushes with short side chains. J. Rheol. **59**(3), 865–883 (2015)
9. S.J. Dalsin, M.A. Hillmyer, F.S. Bates, Molecular weight dependence of zero-shear viscosity in atactic polypropylene bottlebrush polymers. ACS Macro Lett. **3**(5), 423–427 (2014)
10. M. Müllner, Molecular polymer brushes in nanomedicine. Macromol. Chem. Phys. **217**(20), 2209–2222 (2016)
11. R. Verduzco, X. Li, S.L. Pesek, G.E. Stein, Structure, function, self-assembly, and applications of bottlebrush copolymers. Chem. Soc. Rev. **44**(8), 2405–2420 (2015)
12. M. Vatankhah-Varnosfaderani, W.F.M. Daniel, M.H. Everhart, A.A. Pandya, H. Liang, K. Matyjaszewski, A.V. Dobrynin, S.S. Sheiko, Mimicking biological stress–strain behaviour with synthetic elastomers. Nature **549**, 497 (2017)
13. S. Rathgeber, T. Pakula, A. Wilk, K. Matyjaszewski, K.L. Beers, On the shape of bottle-brush macromolecules: systematic variation of architectural parameters. J. Chem. Phys. **122**(12), 124904 (2005)
14. S. Dutta, M.A. Wade, D.J. Walsh, D. Guironnet, S.A. Rogers, C.E. Sing, Dilute solution structure of bottlebrush polymers. Soft Matter **15**(14), 2928–2941 (2019)
15. S.J. Dalsin, M.A. Hillmyer, F.S. Bates, Linear rheology of polyolefin-based bottlebrush polymers. Macromolecules **48**(13), 4680–4691 (2015)
16. L.-H. Cai, T.E. Kodger, R.E. Guerra, A.F. Pegoraro, M. Rubinstein, D.A. Weitz, Soft poly (dimethylsiloxane) elastomers from architecture-driven entanglement free design. Adv. Mater. **27**(35), 5132–5140 (2015)
17. S.L. Pesek, Q. Xiang, B. Hammouda, R. Verduzco, Small-angle neutron scattering analysis of bottlebrush backbone and side chain flexibility. J. Polym. Sci. Part B: Polym. Phys. **55**(1), 104–111 (2017)
18. S.L. Pesek, X. Li, B. Hammouda, K. Hong, R. Verduzco, Small-angle neutron scattering analysis of bottlebrush polymers prepared via grafting-through polymerization. Macromolecules **46**(17), 6998–7005 (2013)
19. F. Snijkers, H.Y. Cho, A. Nese, K. Matyjaszewski, W. Pyckhout-Hintzen, D. Vlassopoulos, Effects of core microstructure on structure and dynamics of star polymer melts: from polymeric to colloidal response. Macromolecules **47**(15), 5347–5356 (2014)
20. A.E. Levi, J. Lequieu, J.D. Horne, M.W. Bates, J.M. Ren, K.T. Delaney, G.H. Fredrickson, C.M. Bates, Miktoarm stars via grafting-through copolymerization: self-assembly and the star-to-bottlebrush transition. Macromolecules **52**(4), 1794–1802 (2019)
21. N.A. Denesyuk, Conformational properties of bottle-brush polymers. Phys. Rev. E Stat. Nonlinear Soft Matter Phys. **67**(5 Pt 1), 051803 (2003)
22. K.J. Bichler, B. Jakobi, S.O. Huber, E.P. Gilbert, G.J. Schneider, Structural analysis of ultrasoft PDMS-g-PDMS shell-only particles. Macromolecules **53**(1), 78–89 (2020)
23. M. Abbasi, L. Faust, K. Riazi, M. Wilhelm, Linear and extensional rheology of model branched polystyrenes: from loosely grafted combs to bottlebrushes. Macromolecules **50**(15), 5964–5977 (2017)
24. D.R. Daniels, T.C.B. McLeish, B.J. Crosby, R.N. Young, C.M. Fernyhough, Molecular rheology of comb polymer melts. 1. Linear viscoelastic response. Macromolecules **34**(20), 7025–7033 (2001)

25. M. Staropoli, A. Raba, C.H. Hövelmann, M.-S. Appavou, J. Allgaier, M. Krutyeva, W. Pyckhout-Hintzen, A. Wischnewski, D. Richter, Melt dynamics of supramolecular comb polymers: viscoelastic and dielectric response. J. Rheol. **61**(6), 1185–1196 (2017)
26. T.C. McLeish, S.T. Milner, Entangled dynamics and melt flow of branched polymers, in *Branched Polymers II*, (Springer, Berlin, 1999), pp. 195–256
27. C. Grigoriadis, A. Nese, K. Matyjaszewski, T. Pakula, H.-J. Butt, G. Floudas, Dynamic homogeneity by architectural design—bottlebrush polymers. Macromol. Chem. Phys. **213** (13), 1311–1320 (2012)
28. J. Pietrasik, B.S. Sumerlin, H.-i. Lee, R.R. Gil, K. Matyjaszewski, Structural mobility of molecular bottle-brushes investigated by NMR relaxation dynamics. Polymer **48**(2), 496–501 (2007)
29. D. Richter, M. Monkenbusch, A. Arbe, J. Colmenero, *Neutron Spin Echo in Polymer Systems* (Springer-Verlag, Berlin, 2005), p. IX-246
30. P.R. Dvornic, J.D. Jovanovic, M.N. Govedarica, On the critical molecular chain length of polydimethylsiloxane. J. Appl. Polym. Sci. **49**(9), 1497–1507 (1993)
31. A.C.C. Esteves, J. Brokken-Zijp, J. Laven, H.P. Huinink, N.J.W. Reuvers, M.P. Van, G. de With, Influence of cross-linker concentration on the cross-linking of PDMS and the network structures formed. Polymer **50**(16), 3955–3966 (2009)
32. B. Jakobi, Silica and siloxane model systems, Dissertation, Louisiana State University, 2019

Chapter 2
Theory

Abstract Theoretical understanding of polymers and characterization methods is required to follow their structural and dynamical behavior. For studying the structural properties, small-angle neutron scattering (SANS) is very suitable due to its available length scale and isotopic sensitivity. The dynamics of polymers range from pico up to nano or milli seconds. This includes fast vibrations, methyl group rotation, segmental dynamics, and polymer dynamics. To cover such a broad time range, a combination of several instruments is needed. Here the concentration is on dielectric spectroscopy and fast field cycling (FFC) relaxometry. Additionally, to get information about the length scales connected with dynamics, quasi-elastic neutron scattering (QENS) was used.

2.1 Polymer

This section introduces the most important aspects of polymers including structure as well as polymer dynamics, important for this dissertation. For further information, it is referred to the respective literature [1–6].

Polymers, also known as macromolecules, are molecules consisting of N chemically identical subunits called repeating units with segment length, ℓ. The average number of monomeric units per macromolecule is known as the degree of polymerization, DP. This defines the molecular weight of the polymer by $M_{\text{polymer}} = \text{DP} \cdot M_{\text{repeating unit}}$ [5].

2.1.1 *Molecular Weight Distribution*

Polymers do not have a single molecular weight, instead, they are showing a molecular weight distribution originating from the polymerization process. The

K.J. Bichler, *Morphology and Dynamics of Bottlebrush Polymers*, Springer Theses,
https://doi.org/10.1007/978-3-030-83379-4_2

molecular weight distribution is characterized by its different moments. Most commonly the number average molecular weight, M_n,

$$M_n := \langle M_n \rangle = \int_0^{\infty} p(M) M \mathrm{d}M \tag{2.1}$$

as the first moment, and the weight average molecular weight, M_w,

$$M_w := \langle M_w \rangle = \frac{\int_0^{\infty} p(M) M^2 \mathrm{d}M}{\int_0^{\infty} p(M) M \, \mathrm{d}M} \tag{2.2}$$

as the second moment of the weight normalized distribution function, $p(M)$, are used [6]. Depending on the polymerization technique different distributions, like Schulz–Zimm or Poison occur. The polydispersity index, PDI $= M_w/M_n$, describes the width of the distribution function and is always greater than or equal to one [7]. A PDI $\cong 1$ with $M_w \cong M_n$ is characteristic of a narrow molecular weight distribution.

2.1.2 Chain Statistics

The simplest approach for describing a polymer chain is the ideal chain model. All segments are linked together, but each of them is randomly oriented independently from the others and the polymer chain can be described by a random walk (Fig. 2.1).

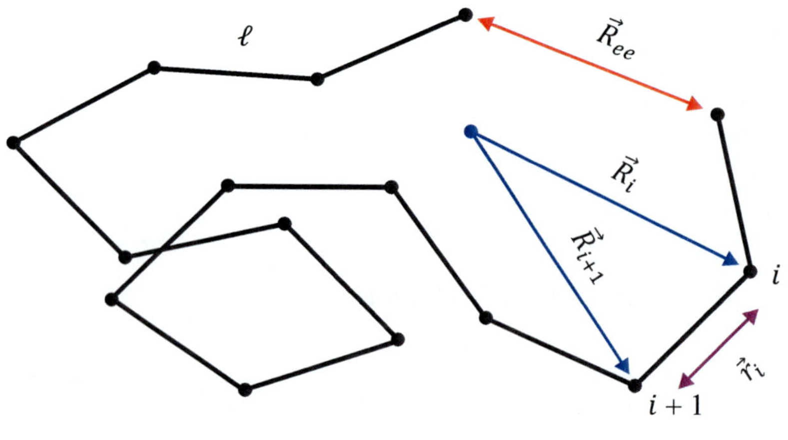

Fig. 2.1 Illustration of an ideal polymer chain

Hereby, the polymer segments are located at positions $\vec{R}_i$ and the distances are given as $\vec{r}_i = \vec{R}_i - \vec{R}_{i+1}$. Ideal chain models typically ignore constraints by finite bonding angles between segments and allow overlap of neighbored chains.

The ensemble average end-to-end vector, $\left\langle \vec{R}_{\mathrm{ee}} \right\rangle$, for the random walk is the sum of all bond vectors in the system and equals zero

$$\left\langle \vec{R}_{\mathrm{ee}} \right\rangle = \sum_{i=1}^{N} \vec{r}_i = 0 \tag{2.3}$$

The simplest non-vanishing moment is the mean square end-to-end vector

$$R_{\mathrm{ee}}^2 := \left\langle \vec{R}_{\mathrm{ee}}^{\,2} \right\rangle = \sum_{i,j=1}^{N} \left\langle \vec{r}_i \cdot \vec{r}_j \right\rangle = \sum_{i=1}^{N} \left\langle \vec{r}_i^{\,2} \right\rangle + \sum_{i \neq j}^{N} \left\langle \vec{r}_i \cdot \vec{r}_j \right\rangle \tag{2.4}$$

Based on the assumption of an ideal chain, there are no correlations between single segments and Eq. (2.4) simplifies to

$$R_{\mathrm{ee}}^2 = \sum_{i=1}^{N} \left\langle \vec{r}_i^{\,2} \right\rangle = N\ell^2 \tag{2.5}$$

In a real polymer, however, the segments cannot order randomly due to geometrical restrictions, like finite and fixed bond angles. In order to transfer the relations from an ideal polymer chain to a real polymer chain, several segments are merged to one segment, called Kuhn segment with the Kuhn length ℓ_{K}. These longer segments are considered to be statistically independent of each other and the approach of an ideal polymer chain becomes valid

$$R_{\mathrm{ee}}^2 = N_{\mathrm{K}} \ell_{\mathrm{K}}^2 \tag{2.6}$$

In order to have a connection with the number of monomeric units, N, the constant C_∞ is introduced. It describes the deviations of the real polymer chain from the ideal chain and is always greater than one. This constant is also known as the Flory characteristic ratio and relates to the local stiffness of a polymer chain

$$R_{\mathrm{ee}}^2 = C_\infty N \ell^2 = N_{\mathrm{K}} \ell_{\mathrm{K}}^2 \tag{2.7}$$

A more general way to estimate the spatial extension of a polymer is given by the mean square radius of gyration, $\left\langle R_{\mathrm{g}}^2 \right\rangle$

$$\left\langle R_g^2 \right\rangle = \frac{1}{N_K} \sum_{i=1}^{N_K} \left(\vec{R}_i - \vec{R}_{cm} \right)^2 \tag{2.8}$$

which is defined as the mean square distance between a Kuhn segment and the center of mass (cm) $\vec{R}_{cm} = 1/N_K \sum \vec{r}_i$. For ideal linear polymer chains, the radius of gyration simplifies to [3, 5, 6]

$$\left\langle R_g^2 \right\rangle = \frac{1}{6} N_K \ell_K^2 = \frac{1}{6} R_{ee}^2 \tag{2.9}$$

For bottlebrush polymers, which have per definition not a Gaussian coil conformation, a description differs from those of linear polymers due to the constraints arising from the repulsive interactions of the side chains. In the spherical region, the overall radius is defined by the length of the side chains and the degree of stretching. For cylindrical bottlebrushes, the radius of the cross-section is given by the length and degree of stretching of the side chains, while the overall cylinder length is determined by the degree of polymerization of the backbone and its degree of stretching. Therefore, radius and length depend on the grafting density. With increasing grafting density, backbone, as well as side chains, are forced to extend more resulting in a larger overall dimension. For low grafting densities, the polymer possesses nearly unperturbed polymer chain statistics. If the grafting density increases, more space is required by side chains belonging to the same backbone. Up to very high grafting densities, the cylinder radius increases with ideal chain statics, $R_{sc} \sim \sqrt{N_{sc}}$, with N_{sc} as the number of monomeric units within one side chain [8].

2.1.3 Dynamics

Depending on the length and time scale, polymer melts show many different processes(Fig. 2.2). The fastest relaxations are fluctuations or vibrations of atoms, followed by jump diffusion processes which could be rotational diffusion of, e.g., methyl groups or small subunits.

These may be rotations of side groups that depend strongly on the local environment, their own shape and size. Segmental relaxation known as α-relaxation follows the vibrational and rotational relaxations in the time scale. It is associated with dynamics of the single chain segments and slowdown with temperature, especially while approaching the glass transition temperature, T_g. To follow this relaxation, dielectric spectroscopy, fast field cycling relaxometry, and neutron scattering have been used in this dissertation to cover a broad frequency/time scale. The global chain dynamics is slower than the segmental relaxation and can be tracked with neutron scattering and depending on the dipole moment also with dielectric spectroscopy. In general, more processes can be observed which are typically referred to as Greek

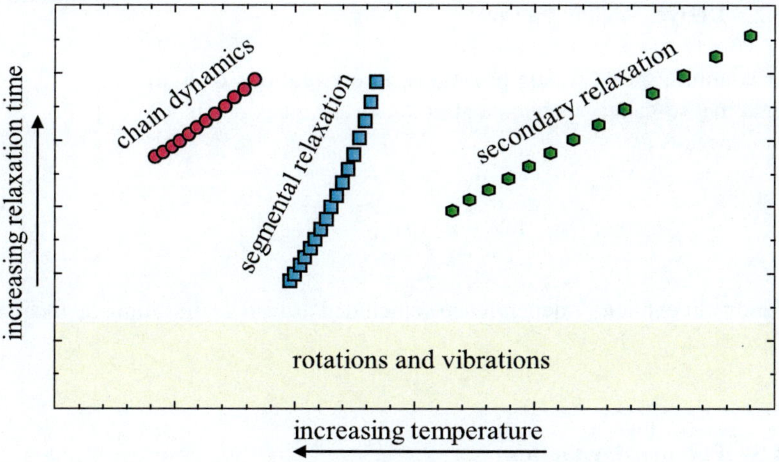

Fig. 2.2 Illustration of the temperature dependence of different relaxation processes typically observed in polymer melts, spread over a broad temperature and time range

letters, like $\beta -$, $\gamma -$, $\delta -$,.... These processes represent fast relaxations at the very local length scale. Typically, they can be observed at low temperatures, i.e., blow T_g.

2.1.3.1 Temperature Dependence

Relaxation processes in polymers show often a temperature dependence that can be described either by the empirical Vogel–Fulcher–Tammann (VFT) equation or the Arrhenius equation.

The VFT equation can be written as:

$$\tau = \tau_\infty \exp\left(\frac{A}{T - T_0}\right) \tag{2.10}$$

The temperature T_0 is the Vogel temperature or ideal glass transition temperature and is usually found to be about $30 - 70$ K below T_g [9]. For high temperatures, τ_∞ describes the limiting relaxation time, i.e., the shortest time in the system and A is a constant [9–11].

For the Arrhenius equation, the universal gas constant, R, and the thermal activation energy, E_A, need to be considered

$$\tau = \tau_\infty \exp\left(\frac{E_A}{RT}\right) \tag{2.11}$$

This law describes thermally activated processes [6, 9].

2.1.3.2 Debye–Waller Factor

For accounting the very fast processes, like vibrations or librations, in the sample system, the so-called Debye–Waller factor is introduced [12, 13]. For isotropic system, it reads as:

$$\mathrm{DWF} = \exp\left(-\frac{\langle u^2\rangle}{3}Q^2\right) \tag{2.12}$$

Hereby an explicit Q-dependence is included as well as the atomistic mean square displacement, $\langle u^2\rangle$.

2.1.3.3 Polymer Dynamics

This dissertation focuses on the segmental relaxation of PDMS-g-PDMS bottlebrush polymers. The polymer dynamics are outside the time window in case of QENS and PDMS has no dipole component parallel to the main chain which would allow to detect the polymer dynamics with dielectric spectroscopy (PDMS is a type B polymer, cf. Sect. 2.3). Therefore, the description of polymer dynamics by Rouse and reptation models is briefly summarized, and only the basic principles, important for this dissertation are described. More detailed information can be found in the literature [2, 6].

If the polymer molecular weight, M, is sufficiently large, the segmental relaxation is almost independent from M. Unlike the α-relaxation, the polymer dynamics relate to the molecular weight. Based on M, polymers can be categorized into two regions—unentangled and entangled—whereby the separation point is known as the entanglement molecular weight, M_e.

For $M < M_e$ the chains can easily move and pass each other. These chains are also known as Rouse chains. For $M > M_e$ the chains are entangled with neighbored ones, which restrict the lateral motions. These chains are reptating chains.

The simplest description of the polymer dynamics of low molecular weight polymers is a bead-spring model, also referred to as the Rouse model. Hereby, a polymer chain is embedded in a heat bath of other chains and consists of N beads, connected by entropic springs (Fig. 2.3). The random fluctuations of the system cause a Brownian motion and interactions with neighboring chains are neglected. For ideal Rouse chains, the longest relaxation time of the system is proportional to the square of the molecular weight, $\tau_R \propto M^2$, known as the Rouse time.

With increasing molecular weight, the polymer chains entangle. The entanglements impose constraints to the single chain dynamics that can be described by a chain relaxation in a tube. The polymer can only reptate inside the infinitely long tube with a diameter, equal to the chain end-to-end distance between two entanglements.

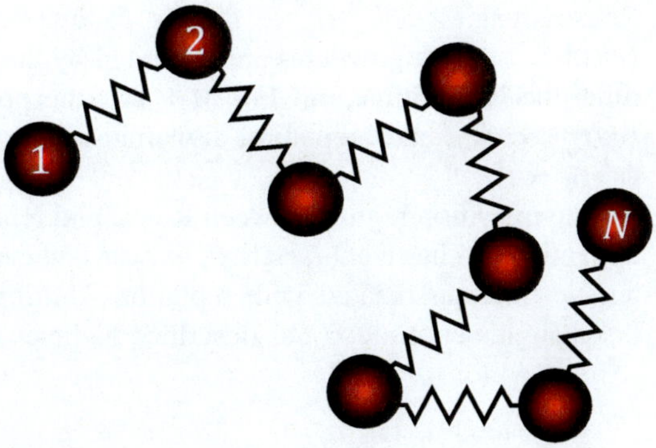

Fig. 2.3 Rouse model represented as a bead spring model with N beads

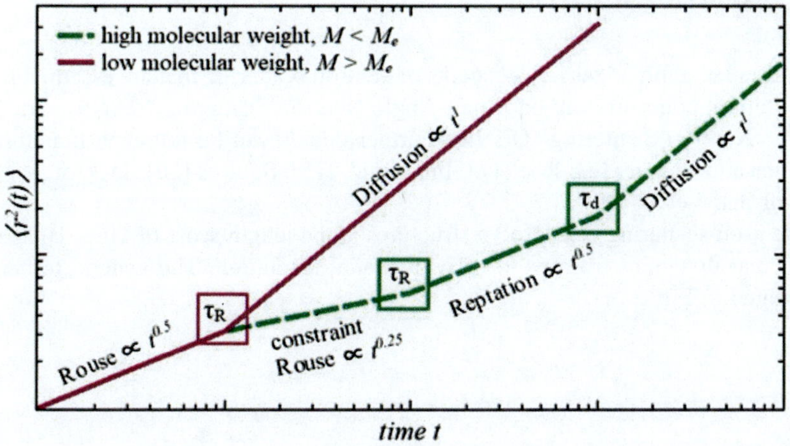

Fig. 2.4 Illustration of the mean square displacement, $\langle r^2(t)\rangle$, vs. time, t, for two different molecular weights

The time needed to escape the tube completely is known as the disentanglement time τ_d and is proportional to M^3, $\tau_d \propto M^3$. Once the polymer is outside the tube, the center of mass diffusion starts, and the behavior is similar to a simple liquid. This description is known as the tube-reptation model. Details on constraint release (CR) and contour length fluctuations (CLF) are omitted.

Both models briefly described above are based on polymers of linear architecture. In this thesis, PDMS based bottlebrush polymers are used, which belong to the family of branched polymers. For this case, the models are more complicated due to the side chains attached to the backbone, which may lead to additional effects, e.g., arm retraction. Therefore, advanced theories are still lacking.

The mean square displacement, $\langle r^2(t)\rangle$, is a common means to describe molecular dynamics, including results from neutron scattering and fast field cycling relaxometry. As illustrated in Fig. 2.4, depending on the molecular weight, different dynamic regions can be identified. For low molecular weights, $M < M_e$, the regime of

Rouse dynamics, $\langle r^2(t)\rangle \propto t^{0.5}$, is followed by the center of mass diffusion, $\langle r^2(t)\rangle \propto t^1$. Both processes are separated by the longest polymer specific relaxation time, the Rouse time, τ_R. For $M > M_e$, the polymer dynamics consist of Rouse, $\langle r^2(t)\rangle \propto t^{0.5}$, and reptation dynamics, $\langle r^2(t)\rangle \propto t^{0.5}$, followed by diffusion, $\langle r^2(t)\rangle \propto t^1$.

The transition region between Rouse and reptation is known as constraint Rouse dynamics and has a characteristic power law dependence of $\langle r^2(t)\rangle \propto t^{0.25}$. Characteristic times associated with reptation additionally to the Rouse time, τ_R, is the disentanglement time, τ_d, as described earlier.

2.2 Neutron Scattering

This chapter summarizes key concepts of neutron scattering that are essential for this dissertation, concentrating on Small-Angle Neutron Scattering (SANS) and Quasi-Elastic Neutron Scattering (QENS). Further details can be found in literature like Higgins and Benoit [12], Roe [14], Imae et al. [15], Squires [16], Jackson [17], Bèe [18] or Sakai et al. [19]

Neutron scattering can resolve structures at the length scale of 10 − 1000 Å and molecular dynamics from ps to ns with spatial resolution. The general principle is illustrated in Fig. 2.5.

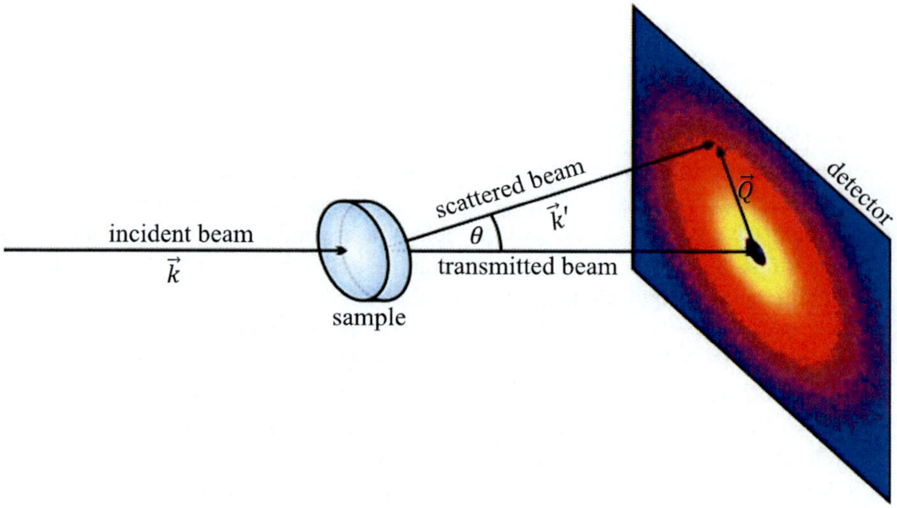

Fig. 2.5 Illustration of a scattering process, including incident beam, $\vec{k}$, scattered beam, $\vec{k}'$, transmitted beam, scattering angle, θ, and momentum transfer, $\vec{Q}$

A neutron hits the sample and is either transmitted, absorbed or scattered under the scattering angle, θ. Thereby, momentum exchanges and the associated momentum transfer, Q, can be written as

$$Q = \left|\vec{Q}\right| = \sqrt{k^2 + k'^2 - 2kk' \cos(2\theta)} \tag{2.13}$$

with $\vec{k}$ and $\vec{k}'$ representing the wavevector of incident and scattered wave and θ the scattering angle. Commonly, a distinction between elastic, inelastic, and quasi-elastic scattering is made. For elastic scattering with $\left|\vec{k}\right| = \left|\vec{k}'\right|$ the relationship between scattering angle and momentum transfer simplifies to

$$Q = \frac{4\pi}{\lambda} \sin\left(\frac{\theta}{2}\right) \tag{2.14}$$

where λ stands for the wavelength.

The whole scattering process can be described by the double differential scattering cross-section, $\frac{d^2\sigma}{d\Omega d\omega}$, i.e., abbreviated as scattering intensity. This is defined as the ratio of scattered neutrons within the surface element, $d\Omega$, having energy transfer, $d\omega$, to the incident neutron flux.

The incoming neutron wave can be treated as a plane wave and the scattering amplitude from a superposition of all scattering centers reads

$$A\left(\vec{Q}, t\right) = \sum_{i=1}^{N} b_i \cdot \exp\left(-i\vec{Q} \cdot \vec{r}(t)\right) \tag{2.15}$$

with b_i the scattering length of atom i. It is not possible to detect the scattering amplitude directly, only the scattering intensity is measured as the absolute square of the scattering amplitude. Therefore, the phase information is lost.

The resulting double differential scattering cross-section has the general form of

$$\begin{aligned} \frac{d^2\sigma}{d\Omega d\omega}\left(\vec{Q}\right) &= \frac{k'}{k}\frac{1}{2\pi} \int \left\langle \left|A\left(\vec{Q}, t\right)\right|^2 \right\rangle \exp(-i\omega t) dt \\ &= \frac{k'}{k}\frac{1}{2\pi} \int \sum_{i,j=1}^{N} \left\langle b_i b_j \cdot \exp\left(-i\vec{Q} \cdot \left(\vec{r}_i(0) - \vec{r}_j(t)\right)\right)\right\rangle \exp(-i\omega t) dt \end{aligned} \tag{2.16}$$

For samples with a distribution of scattering lengths due to a distribution of isotopes and spin states, the scattering length of atoms at positions i and j is uncorrelated

$$\langle b_i b_j \rangle = \begin{cases} \langle b_i \rangle \langle b_j \rangle = \langle b \rangle^2 \equiv b_{\mathrm{coh}}^2 & \text{for } i \neq j \\ \langle b_i b_i \rangle = \langle b^2 \rangle = \langle b \rangle^2 + \left\langle (b - \langle b \rangle)^2 \right\rangle \equiv b_{\mathrm{coh}}^2 + b_{\mathrm{inc}}^2 & \text{for } i = j \end{cases} \tag{2.17}$$

This results in the following double differential scattering cross-section:

$$\begin{aligned} \frac{\mathrm{d}^2\sigma}{\mathrm{d}\Omega \mathrm{d}\omega}\left(\vec{Q}\right) &= \frac{k'}{k}\frac{1}{2\pi} \int \left[\langle b^2 \rangle \sum_{i,j=1}^{N} \left\langle \exp\left(-i\vec{Q} \cdot \left(\vec{r}_i(0) - \vec{r}_j(t)\right)\right) \right\rangle \right. \\ &\quad \left. + \left(\langle b^2 \rangle - \langle b \rangle^2 \right) \sum_{i=1}^{N} \left\langle \exp\left(-i\vec{Q} \cdot \left(\vec{r}_i(0) - \vec{r}_i(t)\right)\right) \right\rangle \right] \exp(-i\omega t) \mathrm{d}t \\ &= N \frac{k'}{k} \left[b_{\mathrm{coh}}^2 S_{\mathrm{coh}}\left(\vec{Q}, \omega\right) + b_{\mathrm{inc}}^2 S_{\mathrm{inc}}\left(\vec{Q}, \omega\right) \right] \end{aligned} \tag{2.18}$$

The first term, also named the coherent scattering function, contains all interference terms and provides insights into the structure. The second part represents the self-correlation of one single particle, known as the incoherent scattering function. For this dissertation only isotropic samples are considered. Therefore, $S\left(\vec{Q}, \omega\right) = S(Q, \omega)$ is used hereafter.

2.2.1 *Small-Angle Neutron Scattering (SANS)*

Small-Angle Neutron Scattering (SANS) measures the elastic scattering, integrated over all energies. This allows to define the differential scattering cross-section, $\mathrm{d}\sigma/\mathrm{d}\Omega$, by integrating over $\mathrm{d}\omega$ and further the macroscopic scattering cross-section, σ, by additional integration over $\mathrm{d}\Omega$

$$\begin{aligned} \frac{\mathrm{d}\sigma}{\mathrm{d}\Omega} &= \int_{-\infty}^{\infty} \frac{\mathrm{d}^2\sigma}{\mathrm{d}\Omega \mathrm{d}\omega} \mathrm{d}\omega \\ \sigma &= \int_{-\infty}^{\infty} \frac{\mathrm{d}\sigma}{\mathrm{d}\Omega} \mathrm{d}\Omega \end{aligned} \tag{2.19}$$

Selected values of the macroscopic scattering cross-section, together with the associated scattering length, b_{coh}, unique for each isotope, are shown in Table 2.1. For SANS, σ_{coh} indicates the strength of the interaction between the neutrons with the scattering center of the object.

In SANS, only the coherent scattering function is considered. The incoherent scattering contribution is subtracted as a constant value. The scattered intensity is usually normalized to the illuminated volume, V, therefore

Table 2.1 Coherent scattering length, b_{coh}, coherent and incoherent scattering cross-section, σ_{coh} and σ_{inc}, for different isotopes used in the samples [20] (1b = 1 barn = 10^{-24} cm^2)

Atom	b_{coh} (fm)	σ_{coh} (b)	σ_{inc} (b)
H	−3.741	1.758	80.27
D	6.671	5.592	2.05
O	5.803	4.232	0.000
Si	4.149	2.163	0.004
C	6.646	5.550	0.001

$$\begin{aligned}\frac{\mathrm{d}\Sigma}{\mathrm{d}\Omega}\left(\vec{Q}\right) &= \frac{1}{V}\frac{\mathrm{d}\sigma}{\mathrm{d}\Omega}\left(\vec{Q}\right) = \frac{1}{V}\left\langle\left|A\left(\vec{Q}\right)\right|^2\right\rangle \\ &= \frac{1}{V}\sum_{i=1}^{N}\left\langle\left|b_i \exp\left(-i\vec{Q}\cdot\vec{r}_i\right)\right|^2\right\rangle\end{aligned} \tag{2.20}$$

By considering a system of N objects, each containing z scattering elements, the coherent scattering cross-section can be rewritten as

$$\begin{aligned}\frac{\mathrm{d}\Sigma}{\mathrm{d}\Omega}\left(\vec{Q}\right) &= \frac{1}{V}\left\langle\left|\sum_{\ell=1}^{N}\exp\left(-i\vec{Q}\cdot\vec{r}_\ell\right)\sum_{i=1}^{z} b_i\exp\left(-i\vec{Q}\cdot\vec{r}_i\right)\right|^2\right\rangle \\ &= \frac{1}{V}\left\langle\left|\sum_{\ell=1}^{N}\exp\left(-i\vec{Q}\cdot\vec{r}_\ell\right)A_\ell\left(\vec{Q}\right)\right|^2\right\rangle \\ &= \frac{1}{V}\left\langle\sum_{\ell=1}^{N}\sum_{m=1}^{N}A_\ell\left(\vec{Q}\right)A_m^*\left(\vec{Q}\right)\exp\left(-i\vec{Q}\cdot\left(\vec{r}_m-\vec{r}_\ell\right)\right)\right\rangle\end{aligned} \tag{2.21}$$

Taking $\ell = m$ as correlated and $\ell \neq m$ as uncorrelated and assuming identical, isotropic objects, Eq. (2.21) can be simplified to

$$\begin{aligned}\frac{\mathrm{d}\Sigma}{\mathrm{d}\Omega}\left(\vec{Q}\right) &= \frac{1}{V} \\ &\times\left[N\left\langle\left|A\left(\vec{Q}\right)\right|^2\right\rangle + \left|\left\langle A\left(\vec{Q}\right)\right\rangle\right|^2\left\langle\sum_{\ell=1}^{N}\sum_{m=1}^{N}\exp\left(-i\vec{Q}\cdot\left(\vec{r}_m-\vec{r}_\ell\right)\right)\right\rangle\right]\end{aligned} \tag{2.22}$$

The introduction of the form factor, $P(Q) = \left\langle\left|A\left(\vec{Q}\right)\right|^2\right\rangle$, and the structure factor, $S(Q) = 1 + \frac{1}{N}\frac{\left|\left\langle A\left(\vec{Q}\right)\right\rangle\right|^2}{\left\langle\left|A\left(\vec{Q}\right)\right|^2\right\rangle}\left\langle\sum_{\ell=1}^{N}\sum_{m=1}^{N}\exp\left(-i\vec{Q}\cdot\left(\vec{r}_m-\vec{r}_\ell\right)\right)\right\rangle$, allows a simplification of Eq. (2.22) to

$$\frac{d\Sigma}{d\Omega}\left(\vec{Q}\right) = \frac{N}{V}P(Q)S(Q) \tag{2.23}$$

The studies here focus on length scales much larger than atomistic dimensions. This allows the definition of the so-called average scattering length density (SLD), $\rho = \frac{\sum_i b_i}{V_m}$, within a volume V_m, containing all scattering elements. This could be, for example, the volume of one repeating unit. The sum of Eq. (2.20) can now be replaced by an integral of the average scattering length density over the whole sample. Therefore, the macroscopic cross-section, normalized to the volume V can be written as

$$\frac{d\Sigma}{d\Omega}\left(\vec{Q}\right) = \frac{1}{V}\left|\int_V \rho\left(\vec{r}\right)\exp\left(-i\vec{Q}\cdot\vec{r}\right)d\vec{r}\right|^2 \tag{2.24}$$

and the scattering amplitude is then a Fourier transformation of the average scattering length density distribution. In general, a two-phase system is very common, i.e., scattering objects with volume V_1 dissolved in a solvent of volume V_2, each phase with a different scattering length density ρ_1 and ρ_2, respectively. Therefore,

$$\rho\left(\vec{r}\right) = \begin{cases} \rho_1 & \text{in } V_1 \\ \rho_2 & \text{in } V_2 \end{cases} \tag{2.25}$$

with total volume $V = V_1 + V_2$. Now Eq. (2.24) can be divided into two subunits, simplified, and written as

$$\begin{aligned} \frac{d\Sigma}{d\Omega}\left(\vec{Q}\right) &= \frac{1}{V}\left|\int_{V_1}\rho_1\exp\left(-i\vec{Q}\cdot\vec{r}\right)d\vec{r}_1 + \int_{V_2}\rho_2\exp\left(-i\vec{Q}\cdot\vec{r}\right)d\vec{r}_2\right|^2 \\ &= \frac{1}{V}\left|\int_{V_1}\rho_1\exp\left(-i\vec{Q}\cdot\vec{r}\right)d\vec{r}_1 + \rho_2\left[\int_V \exp\left(-i\vec{Q}\cdot\vec{r}\right)d\vec{r} - \int_{V_1}\exp\left(-i\vec{Q}\cdot\vec{r}\right)d\vec{r}_1\right]\right|^2 \\ &= \frac{1}{V}(\rho_1-\rho_2)^2\left|\int_{V_1}\exp\left(-i\vec{Q}\cdot\vec{r}\right)d\vec{r}_1\right|^2 \end{aligned} \tag{2.26}$$

The difference in scattering length densities, $\Delta\rho = \rho_1 - \rho_2$, is called contrast and determines the objects visibility for neutrons. Based on the Babinet principle it is irrelevant which scattering length density is greater. Due to the loss of the phase information during Fourier transformation it cannot be determined which value belongs to which phase. The only important value is the difference between both SLDs. In order to create a large contrast between two phases and to be able to resolve the desired structure, isotopic labeling is utilized. Especially the substitution of hydrogen by deuterium has a great influence on the contrast which can be seen on the coherent scattering cross-section, σ_{coh}, from Table 2.1.

2.2.2 Quasi-elastic Neutron Scattering (QENS)

Quasi-Elastic Neutron Scattering (QENS) experiments are associated with energy transfers, $\Delta E = \hbar\omega$ with $\left|\vec{k}\right| \neq \left|\vec{k'}\right|$. This results in information on dynamical behavior, encoded in the energy transfer. In case of the processes studied here, the energy exchanges are in the order of μeV to meV which relates to time scales of ps to ns. This includes motion of molecular reorientation, slow oscillations, and jump processes.

Quasi-elastic neutron scattering can be done by different techniques: (a) time-of-flight spectroscopy and (b) neutron backscattering or (c) a combination of both, in case of a spallation source. The technical principles are slightly different and all have different energy resolutions and ranges, but in general, all measure the intensity depending on the energy transfer, ΔE, for different Q-values. Therefore, the theory explained here is valid for all techniques, unless stated otherwise.

Usually, QENS experiments measure the incoherent scattering function. The incoherent scattering cross-section of hydrogen is much larger compared to other elements and is therefore the dominating part in fully protonated samples. The related experiments are based on incoherent scattering by hydrogen.

In general, a splitting into elastic and quasi-elastic part is possible. Therefore, the measured incoherent scattering function, $S_{inc}(Q, \omega)$, can be written as

$$S_{inc}(Q, \omega) = \mathrm{EISF}(Q)\delta(\omega) + (1 - \mathrm{EISF}(Q))S_{inc}^{qe}(Q, \omega) \tag{2.27}$$

with Q following Eq. (2.13) and ω standing for the energy exchange. Hereby, EISF (Q) is the abbreviation for Elastic Incoherent Structure Factor (EISF) and is mainly included in the elastic line ($\omega = 0$) but contributes partly to the quasi-elastic signal. The EISF provides information on localized motions and on the geometry of the spatial confinement in which the localized process occurs. This can be described with different models, considering different geometries and assumptions.

Here only simple models for the EISF are described. For a rotational threefold jump, common for methyl groups, the EISF can be written as

$$\mathrm{EISF}(Q) = \frac{1}{3}\left(1 + 2j_0\left(QR\sqrt{3}\right)\right) \tag{2.28}$$

with j_0 the zeroth order spherical Bessel function and R being the circles radius. For a diffusive motion within a spherical volume, it results to

$$\mathrm{EISF}(Q) = \left(\frac{3j_1(QR)}{QR}\right)^3 \tag{2.29}$$

with j_1 the first order spherical Bessel function and R the radius of the sphere.

Later the results from several instruments will be used to cover a broader time range compared to a single instrument. It is convenient to Fourier transform the data, which results in the intermediate incoherent scattering function, $S_{\mathrm{inc}}\left(\vec{Q},t\right)$

$$S_{\mathrm{inc}}\left(\vec{Q},t\right) = \frac{1}{N}\sum_i \left\langle \exp\left(i\vec{Q}\cdot\vec{r}_i(t)\right)\cdot\exp\left(-i\vec{Q}\cdot\vec{r}_i(0)\right)\right\rangle \tag{2.30}$$

Usually, $S_{\mathrm{inc}}\left(\vec{Q},t\right)$ represents the self-correlation of objects, whereby $S_{\mathrm{inc}}\left(\vec{Q},t\right) = 0$ means a complete decorrelation, representing a complete relaxation of the system.

To be able to get information about the mean square displacement, $\langle r^2(t)\rangle$, Eq. (2.30) can be expanded to

$$S_{\mathrm{inc}}(Q,t) = A\exp\left(-\frac{\langle r^2(t)\rangle}{6}Q^2 + \frac{\langle r^2(t)\rangle^2\alpha_2(t)}{72}Q^4\right) \tag{2.31}$$

with $\alpha_2(t)$ as the second-order non-Gaussian parameter, which is defined as [21]

$$\alpha_2(t) = \frac{3\langle\Delta r^4(t)\rangle}{5\langle\Delta r^2(t)\rangle^2} - 1 \tag{2.32}$$

2.3 Dielectric Spectroscopy

This chapter provides a summary about the most important aspects to understand the principle of dielectric spectroscopy helpful to understand the presented results. More details can be found in the textbook of Kremer and Schönhals [9].

Dielectric spectroscopy follows the dynamical behavior of a sample under the influence of an external electric field, $\vec{E}(t)$. The dipole moments of the sample align along the electric field and the relaxation of the polarization, $\vec{P}(t)$, is detected [9]. Hereby, two types of polarizations need to be distinguished, orientation and displacement polarization [22, 23]. The first one has the origin in the pure alignment of the permanent dipole moments, $\vec{\mu}$, and provides information about the dynamical behavior. The second one is based on the alignment of induced dipole moments, $\vec{p}$, caused by the electric field and is not used for the determination of dynamical information because the reorientation is too fast. For small $\vec{E}(t)$ the polarization depends linearly on the electric field and the linear response theory is valid. Therefore, the time dependent polarization, $\vec{P}(t)$, as a response of the disturbance by the electric field reads

$$\vec{P}(t) = \epsilon_0 \int_{-\infty}^{t} \epsilon(t - t') \frac{\mathrm{d}\vec{E}(t')}{\mathrm{d}t'} \mathrm{d}t' \tag{2.33}$$

whereby ϵ_0 is the vacuum permittivity constant and $\epsilon(t)$ is the time dependent dielectric function.

For stationary states, with a periodic disturbance, $\vec{E}(t, \omega) = \vec{E}_0 \exp(-i\omega t)$, with ω being the angular frequency, Eq. (2.33) can be rewritten into a simple expression

$$\vec{P}(t, \omega) = \epsilon_0 (\epsilon^*(\omega) - 1) \vec{E}(t, \omega) \tag{2.34}$$

with

$$\epsilon^*(\omega) = \epsilon'(\omega) - i\epsilon''(\omega) \tag{2.35}$$

as the complex dielectric function.

The real part, $\epsilon'(\omega)$, describes the stored energy in the system and the imaginary part, $\epsilon''(\omega)$, is proportional to the energy loss in the system. The relationship between time and frequency domain is given by

$$\epsilon^*(\omega) = \epsilon_\infty - \int_0^\infty \frac{\mathrm{d}\epsilon(t)}{\mathrm{d}t} \exp(-i\omega t) \mathrm{d}t \tag{2.36}$$

with ϵ_∞ as the limiting value for $\epsilon'(\omega \to \infty)$.

Equation (2.36) is a one sided Fourier transformation and therefore, the real and imaginary parts are connected via the Kramers–Kronig relation [9, 23]

$$\begin{aligned} \epsilon'(\omega) &= \epsilon_\infty + \frac{1}{\pi} \oint \frac{\epsilon''(\xi)}{\xi - \omega} \mathrm{d}\xi \\ \epsilon''(\omega) &= -\frac{1}{\pi} \oint \frac{\epsilon'(\xi) - \epsilon_\infty}{\xi - \omega} \mathrm{d}\xi \end{aligned} \tag{2.37}$$

This shows that both quantities contain the same information, if the whole frequency range is known. However, the frequency range, accessible by the experiment is limited and only a relationship between imaginary part and the dielectric strength, $\Delta\epsilon = \epsilon'(\omega = 0) - \epsilon_\infty$, which is represented by the area under the dielectric loss, $\epsilon''(\omega)$, can be determined. This is simply obtained by Eq. (2.36) for $\omega = 0$

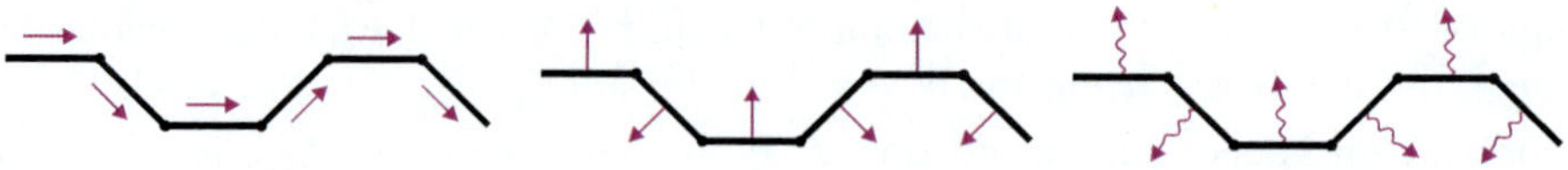

Fig. 2.6 Illustration of the three different permanent dipole moment arrangements based on Stockmayer [24]

$$\Delta\epsilon = \frac{2}{\pi}\int_0^{\infty} \epsilon''(\omega)\mathrm{d}(\ln(\omega)) \tag{2.38}$$

The relaxation processes, detectable with dielectric spectroscopy depend on the orientation of the permanent dipole moment. Hereby, three different types of polymers need to be distinguished as illustrated in Fig. 2.6 [24].

For polymers of type A, the dipole moments are aligned parallel to the backbone and the resulting total polarization is the sum over all permanent dipole moments. In the dynamical sense it represents the fluctuation of the chain end-to-end vector, $\left\langle \vec{R}_{\mathrm{ee}} \right\rangle$, also known as normal mode. This relaxation depends on the molecular weight and the associated relaxation time is the longest in the system.

Polymers of type B possess a permanent dipole moment perpendicular to the main chain. This reflects the segmental mobility due to local changes in the chain conformation. This relaxation is independent of the molecular weight and relates to the glass transition temperature, T_g. It is also known as the α-relaxation. The third type is polymers of type C. Those have a permanent dipole located in the side chain. There is a further relaxation process possible within a polymer which is denoted as a secondary relaxation, named β-process. This process belongs to local fluctuations and side group rotations. In fact, there is no polymer of pure type A available, but polymers having both components, parallel and perpendicular to the main chain, are also referred to as polymers of type A [9]. The polymer focused on in this dissertation, PDMS, is a type B polymer, and only shows the segmental relaxation by dielectric spectroscopy.

2.4 Fast Field Cycling (FFC) Relaxometry

This section summarizes the most important aspects of fast field cycling relaxometry used for this work. More details can be found in the literature, e.g., Kimmich and Anoardo [25], Kimmich [26], Fujara et al. [27] or Anoardo et al. [28]

Fast field cycling relaxometry measures the Larmor frequency dependence of the spin–lattice relaxation rate, $R_1(\omega) = \frac{1}{T_1(\omega)}$, by rapidly switching (cycling) the external magnetic detection field, whereby $T_1(\omega)$ stands for the spin–lattice relaxation time. Here $\omega = -\gamma_H B_0$ is the Larmor frequency with γ_H the gyromagnetic ratio of the protons and B_0 the experimentally controlled external magnetic field, the sample is exposed to.

For nuclei with spin $\frac{1}{2}$, the main interaction controlling the proton spin–lattice relaxation is the magnetic dipole–dipole interaction of different protons in the system, with the Hamiltonian [29, 30]

$$\widehat{H}_{\mathrm{DD}}(t) = \sum_{i<j} \frac{\mu_0}{4\pi} \frac{\gamma_{\mathrm{H}}^2 \hbar^2}{r_{ij}^3} \left\{ \widehat{\vec{I}}_i \cdot \widehat{\vec{I}}_j - 3\left(\widehat{\vec{I}}_i \cdot \vec{e}_{ij}(t) \right) \left(\widehat{\vec{I}}_j \cdot \vec{e}_{ij}(t) \right) \right\} \tag{2.39}$$

including r_{ij} the internuclear distance, $\vec{e}_{ij} = \frac{\vec{r}_{ij}}{r_{ij}}$ the unity vector, $\widehat{\vec{I}}_{i/j}$ the spin vector operator of spin number i/j, and μ_0 the magnetic field constant. Here only interactions involving two spins are considered. The time dependency is due to fluctuations of the magnetic field generated by neighboring spins. In contrast, for spin 1 nuclei quadrupole coupling needs to be considered as the dominate part [25].

The relationship with dynamics is obtained from the second-order time dependent perturbation theory and the resulting relaxation rates can be expressed in the Redfield-limit, i.e., $\omega T_1 \gg 1$, as follows [29]

$$R_1(\omega) = \left(\frac{\mu_0}{4\pi}\right)^2 \frac{6\pi}{5} \gamma_{\mathrm{H}}^4 \hbar^2 \int_0^\infty \{\cos(\omega t) A_1(t) + 4\cos(2\omega t) A_2(t)\} \mathrm{d}t \tag{2.40}$$

This is a linear combination of Fourier transformations of the time dependent dipolar correlation function $A_m(t)$ of second rank ($l = 2$) with $A_m(t)$ written as [29, 31]

$$A_m(t) = \frac{1}{N_{\mathrm{s}}} \sum_{i \neq j} \left\langle \frac{Y_{2,m}\left(\vec{e}_{ij}(t)\right)}{r_{ij}^3(t)} \frac{Y_{2,m}^*\left(\vec{e}_{ij}(0)\right)}{r_{ij}^3(0)} \right\rangle \tag{2.41}$$

$Y_{lm}\left(\vec{e}_{ij}\right)$ is the spherical harmonic of rank $l = 2$ and order m, and N_{s} describes the total number of spins in the system. For isotropic systems like a polymer melt, the correlation function is independent of m, i.e., $A_{m=0}(t) = A_1(t) = A_2(t) \equiv A(t)$ and Eq. (2.41) reduces to the even correlation function, containing the second rank Legendre polynomial with the angle, $\theta(t)$, specifying the orientation of the internuclear vector

$$A(t) = \frac{5}{16\pi N_{\mathrm{s}}} \sum_{i \neq j} \left\langle \frac{3\cos^2(\theta(t)) - 1}{r_{ij}^3(t)} \frac{3\cos^2(\theta(0)) - 1}{r_{ij}^3(0)} \right\rangle \tag{2.42}$$

This correlation function includes intra- and intermolecular contributions, i.e., correlations between spins belonging to the same macromolecule and between spins from different macromolecules. Therefore, the correlation function can be decomposed into a sum of intra- and intermolecular contributions

$$A(t) = A^{\text{intra}}(t) + A^{\text{inter}}(t) \tag{2.43}$$

Based on the Wiener/Khinchine theorem, the Fourier transform of the autocorrelation function represents the spectral density. Therefore, the dipolar spectral density, $J_{\text{DD}}(\omega)$, of the even correlation function, $A(t)$, can be written as [25, 26, 32]

$$J_{\text{DD}}(\omega) = \frac{1}{2}\int_{-\infty}^{\infty} A(t)\exp(-i\omega t)\text{d}t = \int_{0}^{\infty} A(t)\cos(\omega t)\text{d}t \tag{2.44}$$

Using Eq. (2.44) allows to rewrite Eq. (2.40) into the Bloembergen–Purcell–Pound (BPP) equation [25, 26]

$$R_1(\omega) = K \cdot [J_{\text{DD}}(\omega) + 4J_{\text{DD}}(2\omega)] \tag{2.45}$$

whereby $K = \left(\frac{\mu_0}{4\pi}\right)^2 \frac{6\pi}{5}\gamma_{\text{H}}^4\hbar^2$ is the dipolar coupling constant. Multiplication of the spectral density with ω yields the dynamic susceptibility, $\chi''(\omega) = \omega J_{\text{DD}}(\omega)$, and allows to rewrite the BPP equation to

$$\omega R_1(\omega) = K \cdot [\chi''(\omega) + 2\chi''(2\omega)] \cong 3K\chi''_{\text{FC}}(\omega) \tag{2.46}$$

whereby $\chi''(\omega) \cong \chi''(2\omega)$ results in a total field cycling (FC) susceptibility, $\chi''_{\text{FC}}(\omega)$.

With the assumption that the NMR coupling constant K does not change significantly with the temperature, the dipolar susceptibility, $\chi''_{\text{DD}}(\omega)$, can be introduced, allowing to obtain susceptibility master curves

$$\omega R_1(\omega) \cong 3K\chi''_{\text{FC}}(\omega) = \chi''_{\text{DD}}(\omega) \tag{2.47}$$

This permits to extend the narrow frequency range of FFC, by measuring at different temperatures and utilizing the Time–Temperature Superposition principle (TTS). Similar to the correlation function, $A(t)$, the susceptibility, $\chi''_{\text{DD}}(\omega)$, and the relaxation rate, $R_1(\omega)$, consist also of the two contributions, intra- and intermolecular contribution (Fig. 2.7).

$$\begin{aligned} \chi''_{\text{DD}}(t) &= \chi''_{\text{intra}}(t) + \chi''_{\text{inter}}(t) \\ R_1(\omega) &= R_1^{\text{intra}}(\omega) + R_1^{\text{inter}}(\omega) \end{aligned} \tag{2.48}$$

Usually, isotopic dilution experiments are needed to separate the intermolecular relaxation, χ''_{inter}, from the measured susceptibility, χ''_{DD}. There are only a few exceptions, e.g., PDMS [33, 34].

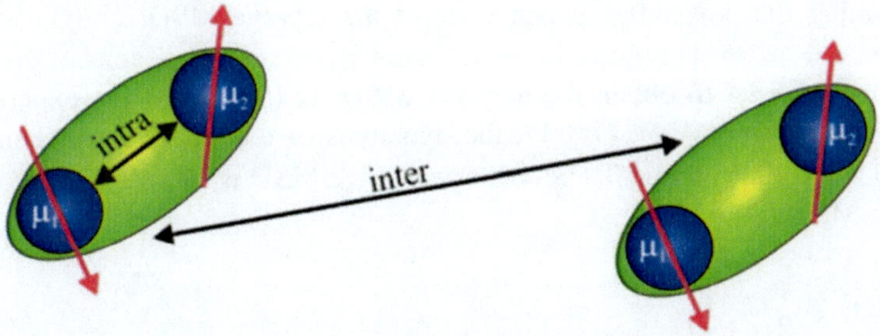

Fig. 2.7 Illustration of intra- and intermolecular interaction of molecules with nuclear spin pairs

Using the intermolecular dynamics further, a mean square displacement, $\langle r^2(t)\rangle$, can be generated. Therefore, the relaxation rate, obtained from perturbation theory, for the intermolecular part only, can be written similar to Eq. (2.40).

In order to transform the relaxation rate into the time domain, Eq. (2.40) can be rewritten by an inverse Fourier transformation to [29, 34]

$$A^{\text{inter}}(t) + 2A^{\text{inter}}\left(\frac{t}{2}\right) = \left(\frac{4\pi}{\mu_0}\right)^2 \frac{5}{3\pi^2} \frac{1}{\gamma_H^4 \hbar^2} \int_0^\infty \frac{\cos(\omega t)}{T_1^{\text{inter}}(\omega)} d\omega \tag{2.49}$$

The intermolecular contribution of the correlation function is given in a universal form. It is independent of the model for polymer dynamics and can be written as

$$A^{\text{inter}}(t) = \sqrt{\frac{2}{3\pi}} \frac{N_s}{\left(\langle \tilde{r}^2(t)\rangle^{1/2}\right)^3} \tag{2.50}$$

with $N_s = n_H \frac{\rho N_A}{M}$ denoting the spin density [35] which can be calculated based on the number of hydrogen atoms per monomeric unit, n_H, mass density, ρ, molar mass, M, of one monomeric unit and the Avogadro constant, N_A.

Substituting $A^{\text{inter}}(t)$ (Eq. 2.50) in Eq. (2.49) introduces a relationship between the intermolecular relaxation and the relative mean square displacement, $\langle \tilde{r}^2(t)\rangle$

$$\frac{1}{\langle \tilde{r}^2(t)\rangle^{\frac{3}{2}}} + \frac{2}{\langle \tilde{r}^2(t/2)\rangle^{\frac{3}{2}}} = \frac{5}{4N_s} \sqrt{\frac{8}{3\pi^3}} \left(\frac{4\pi}{\mu_0}\right)^2 \frac{1}{\gamma_H^4 \hbar^2} \int_0^\infty \frac{\cos(\omega t)}{T_1^{\text{inter}}(\omega)} d\omega \tag{2.51}$$

The left-hand side of Eq. (2.51) can be simplified to

$$\frac{1}{\langle \tilde{r}^2(t)\rangle^{\frac{3}{2}}} + \frac{2}{\langle \tilde{r}^2(t/2)\rangle^{\frac{3}{2}}} = \frac{1 + 2^{1+\frac{3\alpha}{2}}}{\langle \tilde{r}^2(t)\rangle^{\frac{3}{2}}} \tag{2.52}$$

by assuming the power law dependence of the relative MSD, $\langle \tilde{r}^2(t) \rangle \propto t^\alpha$, with exponent $\alpha = [0.25; 1]$ typical for the different characteristic regions of the polymer dynamics. In order to obtain the absolute MSD, $\langle r^2(t) \rangle$, out of the experimentally determined relative MSD, $\langle \tilde{r}^2(t) \rangle$, the dynamics of the chains are assumed to be independent of each other. Therefore, the relative MSD is estimated to be twice the absolute MSD [29, 34]

$$\begin{aligned} \langle r^2(t) \rangle &= \frac{1}{2} \langle \tilde{r}^2(t) \rangle \\ &= \frac{1}{2} \left[\frac{1}{1 + 2^{1+\frac{3\alpha}{2}}} \frac{5}{4N_\mathrm{s}} \sqrt{\frac{8}{3\pi^3}} \left(\frac{4\pi}{\mu_0} \right)^2 \frac{1}{\gamma_\mathrm{H}^4 \hbar^2} \int_0^\infty \frac{\cos(\omega t)}{T_1^\mathrm{inter}(\omega)} \mathrm{d}\omega \right]^{-\frac{2}{3}} \end{aligned} \tag{2.53}$$

In the low frequency region, the intermolecular relaxation contribution decays slowly for all polymer models and allows to determine the diffusion coefficient, D. Therefore, the intermolecular contribution can be rewritten as [29]

$$\begin{aligned} R_1^\mathrm{inter}(\omega) &= \left(\frac{\mu_0}{4\pi} \right)^2 \frac{4}{5} \frac{\sqrt{3\pi}}{2} N_\mathrm{s} \gamma_\mathrm{H}^4 \hbar^2 \int_0^\infty \frac{[\cos(\omega t) + 4\cos(2\omega t)]}{\langle \tilde{r}^2(t) \rangle^{\frac{3}{2}}} \mathrm{d}t \\ &= \left(\frac{\mu_0}{4\pi} \right)^2 \frac{4}{5} \frac{\sqrt{3\pi}}{2} N_\mathrm{s} \gamma_\mathrm{H}^4 \hbar^2 \cdot \int_0^\infty \left(\frac{1}{\langle \tilde{r}^2(t) \rangle^{\frac{3}{2}}} + \frac{2}{\langle \tilde{r}^2(t/2) \rangle^{\frac{3}{2}}} \right) \left(1 - 2\sin^2\left(\frac{\omega t}{2} \right) \right) \mathrm{d}t \end{aligned} \tag{2.54}$$

For the limiting case of $\omega\tau_\mathrm{d} \to 0$ with τ_d being the terminal relaxation time, i.e., the longest relaxation time in a polymer melt, (for systems showing Rouse dynamics $\tau_\mathrm{d} = \tau_\mathrm{R}$ with τ_R as the Rouse time) the frequency dependence of Eq. (2.54) vanishes and the relaxation rate, $R_1^\mathrm{inter}(0)$, reaches a plateau value. Using

$$\langle r^2(t) \rangle \cong 6Dt + 2R_\mathrm{g}^2 \tag{2.55}$$

which is valid for $t \gg \tau_\mathrm{d}$ and originated in the Rouse model with R_g the radius of gyration, the lower limiting value for the relaxation rate is obtained

$$R_1^\mathrm{inter}(0) \geq \left(\frac{\mu_0}{4\pi} \right)^2 \sqrt{\frac{\pi}{6}} \left(\frac{\gamma_\mathrm{h}^4 \hbar^2 N_\mathrm{s}}{R_\mathrm{g} D} \right) \tag{2.56}$$

Applying a correction to the plateau value by using Eqs. (2.54) and (2.55), neglecting R_g a universal dispersion relation can be extracted [25, 29, 34]

$$R_1^{\text{inter}}(\omega) = R_1^{\text{inter}}(0) - \left(\frac{\mu_0}{4\pi}\right)^2 \frac{\pi}{30}\left(1 + 4\sqrt{2}\right)\gamma_{\text{h}}^4 \hbar^2 N_{\text{s}} \frac{\sqrt{\omega}}{D^{3/2}} \tag{2.57}$$

Once the center of mass diffusion coefficient is known, the radius of gyration, R_g, can be extracted by using Eq. (2.56).

References

1. P.J. Flory, *Principles of Polymer Chemistry* (Cornell University Press, Ithaca, 1953)
2. M. Doi, S.F. Edwards, *The Theory of Polymer Dynamics* (Clarendon Press, Oxford, 1988)
3. M. Doi, H. See, *Introduction to Polymer Physics* (Clarendon Press, Oxford, 1996)
4. P.G. de Gennes, *Introduction to Polymer Dynamics* (Cambridge University Press, Cambridge, 1990)
5. M. Rubinstein, R.H. Colby, *Polymer Physics* (OUP, Oxford, 2003)
6. G.R. Strobl, *The Physics of Polymers: Concepts for Understanding Their Structures and Behavior* (Springer, Berlin, 2013)
7. H.G. Elias, *Makromoleküle: Chemische Struktur und Synthesen - Sechste, vollstandig überarbeitete Auflage* (Wiley, New York, 2009)
8. W.F. Daniel, J. Burdynska, M. Vatankhah-Varnoosfaderani, K. Matyjaszewski, J. Paturej, M. Rubinstein, A.V. Dobrynin, S.S. Sheiko, Solvent-free, supersoft and superelastic bottlebrush melts and networks. Nat. Mater. **15**(2), 183–189 (2016)
9. F. Kremer, A. Schönhals, *Broadband Dielectric Spectroscopy* (Springer, Berlin, 2012)
10. D.I. Bower, *An Introduction to Polymer Physics* (Cambridge University Press, Cambridge, 2002)
11. R. Lund, A. Alegria, L. Goitandia, J. Colmenero, M.A. González, P. Lindner, Dynamical and structural aspects of the cold crystallization of poly(dimethylsiloxane) (PDMS). Macromolecules **41**(4), 1364–1376 (2008)
12. J.S. Higgins, H.C. Benoit, H. Benoît, *Polymers and Neutron Scattering* (Clarendon Press, Oxford, 1996)
13. D. Richter, M. Monkenbusch, A. Arbe, J. Colmenero, *Neutron Spin Echo in Polymer Systems* (Springer-Verlag, Berlin, 2005), p. IX-246
14. P.M.S.R.J. Roe, R.J. Roe, *Methods of X-ray and Neutron Scattering in Polymer Science* (Oxford University Press, Oxford, 2000)
15. T. Imae, T. Kanaya, M. Furusaka, N. Torikai, *Neutrons in Soft Matter* (Wiley, New York, 2011)
16. G.L. Squires, *Introduction to the Theory of Thermal Neutron Scattering* (Cambridge University Press, Cambridge, 2012)
17. A.J. Jackson, *Introduction to Small-Angle Neutron Scattering and Neutron Reflectometry* (NIST Center for Neutron Research, Gaithersburg, 2008), pp. 1–24
18. M. Bée, *Quasielastic Neutron Scattering* (Adam Hilger, Bristol, 1988)
19. V.G. Sakai, C. Alba-Simionesco, S.H. Chen, *Dynamics of Soft Matter: Neutron Applications* (Springer US, New York, 2011)
20. A.-J. Dianoux, G. Lander, *Neutron Data Booklet* (OCP Science, Philadelphia, 2003)
21. C. Gerstl, G.J. Schneider, A. Fuxman, M. Zamponi, B. Frick, T. Seydel, M. Koza, A.C. Genix, J. Allgaier, D. Richter, J. Colmenero, A. Arbe, Quasielastic neutron scattering study on the dynamics of poly(alkylene oxide)s. Macromolecules **45**(10), 4394–4405 (2012)
22. C. Gerthsen, D. Meschede, *Gerthsen Physik* (Springer, Berlin, 2006)
23. D. Haarer, H.W. Spiess, *Spektroskopie Amorpher und Kristalliner Festkörper* (Steinkopff, Darmstadt, 2013)

24. W.H. Stockmayer, Dielectric dispersion in solutions of flexible polymers. Pure Appl. Chem. **15** (3–4), 539–554 (1967)
25. R. Kimmich, E. Anoardo, Field-cycling NMR relaxometry. Prog. Nucl. Magn. Reson. Spectrosc. **44**(3), 257–320 (2004)
26. R. Kimmich, *Field-Cycling NMR Relaxometry: Instrumentation. Model Theories and Applications* (Royal Society of Chemistry, London, 2018)
27. F. Fujara, D. Kruk, A.F. Privalov, Solid state field-cycling NMR relaxometry: instrumental improvements and new applications. Prog. Nucl. Magn. Reson. Spectrosc. **82**, 39–69 (2014)
28. E. Anoardo, G. Galli, G. Ferrante, Fast-field-cycling NMR: applications and instrumentation. Appl. Magn. Reson. **20**(3), 365–404 (2001)
29. N. Fatkullin, S. Stapf, M. Hofmann, R. Meier, E.A. Rossler, Proton spin dynamics in polymer melts: new perspectives for experimental investigations of polymer dynamics. J. Non·Cryst. Solids **407**, 309–317 (2015)
30. P.A. Abragam, A. Abragam, *The Principles of Nuclear Magnetism* (Clarendon Press, Oxford, 1961)
31. R. Kimmich, N. Fatkullin, Self-diffusion studies by intra- and inter-molecular spin-lattice relaxometry using field-cycling: liquids, plastic crystals, porous media, and polymer segments. Prog. Nucl. Magn. Reson. Spectrosc. **101**, 18–50 (2017)
32. A. Khintchine, Korrelationstheorie der Stationären Stochastischen Prozesse. Math. Ann. **109** (1), 604–615 (1934)
33. A. Herrmann, B. Kresse, M. Wohlfahrt, I. Bauer, A.F. Privalov, D. Kruk, N. Fatkullin, F. Fujara, E.A. Rössler, Mean square displacement and reorientational correlation function in entangled polymer melts revealed by field cycling 1H and 2H NMR relaxometry. Macromolecules **45**(16), 6516–6526 (2012)
34. B. Kresse, M. Hofmann, A.F. Privalov, N. Fatkullin, F. Fujara, E.A. Rössler, All polymer diffusion regimes covered by combining field-cycling and field-gradient 1H NMR. Macromolecules **48**(13), 4491–4502 (2015)
35. R. Meier, A. Herrmann, M. Hofmann, B. Schmidtke, B. Kresse, A.F. Privalov, D. Kruk, F. Fujara, E.A. Rössler, Iso-frictional mass dependence of diffusion of polymer melts revealed by 1H NMR relaxometry. Macromolecules **46**(14), 5538–5548 (2013)

Chapter 3
Experimental Techniques

Abstract Information about the structural and dynamical behavior of bottlebrush polymers can be obtained with different techniques. Hereby, small-angle neutron scattering (SANS) has been used to investigate the structure in melt as well as in solution while taking advantage of the neutron's isotopic sensitivity. For the dynamical part, several instruments have been used to cover the relaxation behavior of bottlebrush polymers in the time range from pico to nano seconds. The segmental relaxation of poly(dimethylsiloxane)-based polymers can be tracked well by dielectric spectroscopy over several temperatures and can additionally be extended to longer relaxation times by using fast field cycling (FFC) relaxometry. Additionally, FFC relaxometry offers access to large scale polymer dynamics by means of the time dependent mean square displacement. To get information about dynamical processes depending on their respective length scale, quasi-elastic neutron scattering (QENS) was the method of choice, having access to fast vibrations, methyl group rotation, and segmental dynamics.

This chapter briefly describes the principles of the used instruments. Further information about the devices, especially about the neutron scattering instruments can be found in the respective instrument publications, QUOKKA [1, 2], KWS-2 [3], BILBY [4, 5], EMU [6], PELICAN [7], IRIS [8], and in related literature [9–11].

3.1 Small-Angle Neutron Scattering (SANS)

Small-Angle Neutron Scattering (SANS) was used to determine the morphology of the PDMS based bottlebrush polymers in different environments. A broad Q-range allows the determination of structural features, starting from the overall shape to the arrangements within the bottlebrush shell. Three different instruments have been used, whereby the temperature was $T = 21$ °C for all measurements. In all cases, the samples were placed in Hellma cells with a thickness of $d = 1$ mm, ensuring high transmission and minimizing multiple scattering effects.

K.J. Bichler, *Morphology and Dynamics of Bottlebrush Polymers*, Springer Theses,
https://doi.org/10.1007/978-3-030-83379-4_3

3.1.1 QUOKKA and KWS-2: Pinhole SANS

The QUOKKA instrument is located at the ANSTO facility in Sydney, Australia and the KWS-2 instrument is at the FRM II in Munich, Germany. Both are cold-neutron instruments with a pinhole setup whereby the neutron beam is focused on the sample by different apertures (Fig. 3.1).

The incoming neutrons are monochromatized by a velocity selector, yielding a wavelength distribution of $\frac{\Delta\lambda}{\lambda} = 10\%$ for the wavelength $\lambda = 5$ Å. Before the sample, the neutron beam is collimated and after scattering at the sample, neutrons are detected at the 2D detector. To be able to resolve length scales from a few nanometers to almost micrometer, different sample-to-detector distances are necessary. Usually, different sample-to-detector distances are used (Table 3.1), resulting in a Q-range of $0.003\ \text{Å}^{-1} < Q < 0.5\ \text{Å}^{-1}$ (KWS-2) and $0.003\ \text{Å}^{-1} < Q < 0.8\ \text{Å}^{-1}$ (QUOKKA). In order to increase the maximum accessible Q-range for QUOKKA, the sample-to-detector distance of $L_1 = 1.3$ m was connected with a 300 mm lateral detector offset. These configurations yield length scales, d, from $7.85\ \text{Å} < d < 2094\ \text{Å}$.

The data reduction was performed by using the suite of NCNR SANS reduction macros [12], modified for the QUOKKA instrument, in the IGOR software package (WaveMetrics, Portland, OR) or for measurements at KWS-2 with the QtiKWS software together with suitable routines. After correction for empty cell scattering, transmission, and detector response, the data were radially averaged and transformed onto an absolute scale by calibration with an attenuated direct beam (QUOKKA) or with the secondary standard poly(methylmethacrylate), PMMA (KWS-2). All samples were additionally background corrected by subtracting the corresponding contributing volume fraction from the linear PDMS melt or solvent and incoherent scattering contribution.

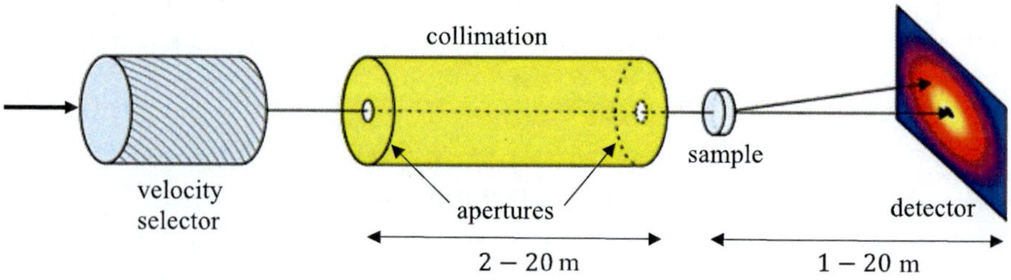

Fig. 3.1 Illustration of a small-angle neutron scattering beamline, including velocity selector, apertures for collimation, sample, and the detector. The whole setup is within a vacuum vessel

Table 3.1 Sample-to-detector distances for the two different instruments used for the experiments

Instrument	L_1 (m)	L_2 (m)	L_3 (m)
QUOKKA	1.3	8	20
KWS-2	2	8	20

3.1.2 *BILBY: Time-of-Flight SANS*

The BILBY instrument is also located at the ANSTO facility in Sydney, Australia, but unlike the QUOKKA instrument it is a time-of-flight based SANS instrument, able to work in two different modes. Monochromatic mode like QUOKKA with a velocity selector, filtering neutrons of a defined wavelength, λ=6 Å and resolution $\frac{\Delta\lambda}{\lambda} = 10\%$ and in the time-of-flight mode, whereby the velocity selector is replaced by four disk choppers, which creates neutron pulses of adjustable resolution $\frac{\Delta\lambda}{\lambda} = 3.0 - 30\%$ (Fig. 3.2). To focus the neutron beam on the sample, the pinhole setup is used, similar to QUOKKA. For this dissertation, the monochromatic mode was used.

One advantage of BILBY is the use of two detector carriers which can move independently within the vacuum vessel. The detector carriers consist of one single rear detector and a complex front detector consisting of four equal panels, called curtains. Hereby, distances from the sample to the rear detector, the horizontal and the vertical curtains of 13.491, 1.893, and 0.889 m were used, whereby the separation of the left and right curtains was 0.175 and 0.350 m and those of the top and bottom curtain was 0.150 and 0.030 m (Fig. 3.3). This allows to capture the whole

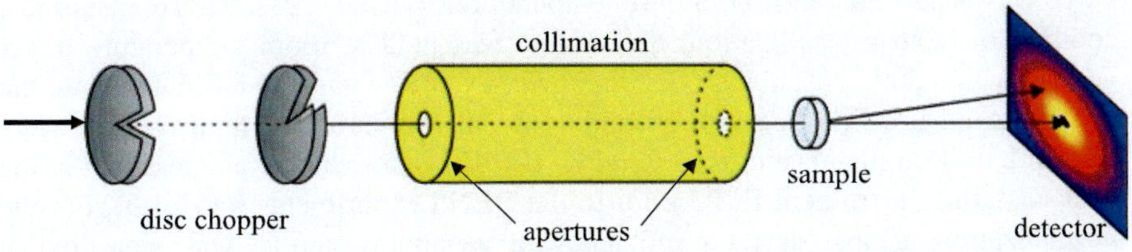

Fig. 3.2 Illustration of a time-of-flight small-angle neutron scattering beamline, including disc chopper, apertures for collimation, sample, and a simplified detector. The whole setup is within a vacuum vessel

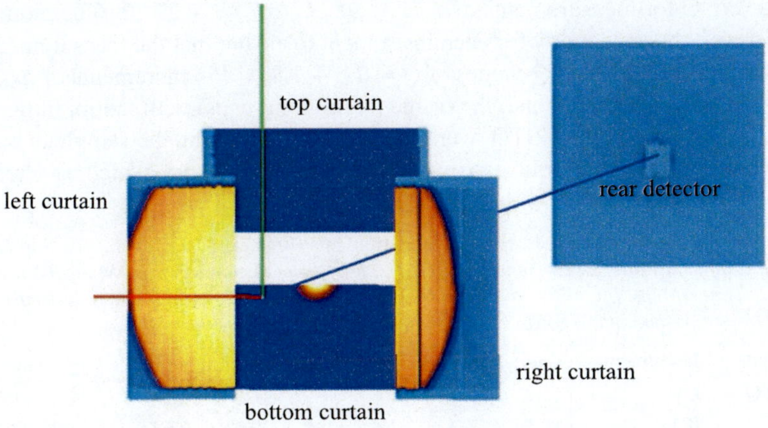

Fig. 3.3 Schematic representation of the detector setup of the BILBY beamline

Q-range, 0.003 $\mathring{A}^{-1} < Q < 0.56\ \mathring{A}^{-1}$, within one measurement, without changing the sample-to-detector distance. The data reduction was done similar to QUOKKA by using MANTID instead of IGOR.

3.2 Quasi-elastic Neutron Scattering (QENS)

Quasi-Elastic Neutron Scattering (QENS) was used to determine the dynamical behavior of bottlebrush polymers with spatial resolution. In order to cover a broad frequency/time range, three different instruments were combined. Each of them having a different energy transfer and energy resolution, resulting in different time ranges. The technical specifications of the instruments regarding their energy resolution and energy range are summarized in Table 3.2.

All measurements were conducted in a temperature range of $T = -198$ °C to $T = +27$ °C, whereby the instrument resolution was measured at $T = -271.5$ °C for EMU and PELICAN and at $T < -263$ °C for IRIS. For the temperature control, a closed cycle cryostat with secondary helium circulation loop, controlled by a needle valve was used. This allowed a base temperature of $T < -263$ °C. To ensure fast cooling to avoid crystallization, the sample was held at room temperature under vacuum while the cryostat was cooled. Finally exchange gas was introduced into the system to ensure fast quenching condition. All measurement temperatures were chosen based on the elastic fixed window (EFW) scan, which was measured at the backscattering instrument EMU prior to the QENS experiments (cf. 5.1.3). For the normalization against detector efficiency, a vanadium sample was measured at $T = +27$ °C. Additionally an empty can was measured at the lowest, $T = -198$ °C, and at the highest, $T = +27$ °C, temperature and subtracted from the measurement signal. Hereby the empty can of $T = -198$ °C was used for measurements in the temperature range of $T = -198$ °C to $T = -123$ °C and the empty can of $T = +27$ °C for measurements of $T = -98$ °C to $T = +27$ °C. The momentum transfer was chosen equally for each instrument to ensure that the measurements can be combined, resulting in Q-values of $Q = 0.6 - 1.8\ \mathring{A}^{-1}$ in increments of $\Delta Q = 0.2\ \mathring{A}^{-1}$. For all measurements, the same aluminum cans with aluminum screws (Fig. 3.4), provided by ANSTO, were used without removing the sample in between the beamtimes. The sample gap was chosen to be 0.1 mm which ensures 90%

Table 3.2 Specifications of the used QENS instruments *(Bichler, K. J., Jakobi, B., García Sakai, V., Klapproth, A., Mole, R. A., Schneider, G. J., Short Time Dynamics of PDMS-g-PDMS Bottle-brush Polymer Melts Investigated by Quasi-Elastic Neutron Scattering. Macromolecules 2020, 10.1021/acs.macromol.0c01846)*

Location	Instrument	Energy resolution (μeV)	Energy range (meV)	Q-range ($\mathring{A}^{-1}$)
ANSTO	EMU	1.1	−0.028 to 0.028	0.4 – 1.8
ISIS	IRIS	17.5	−0.55 to 0.57	0.6 – 1.6
ANSTO	PELICAN	65	−19.99 to 2.59	0.2 – 1.8

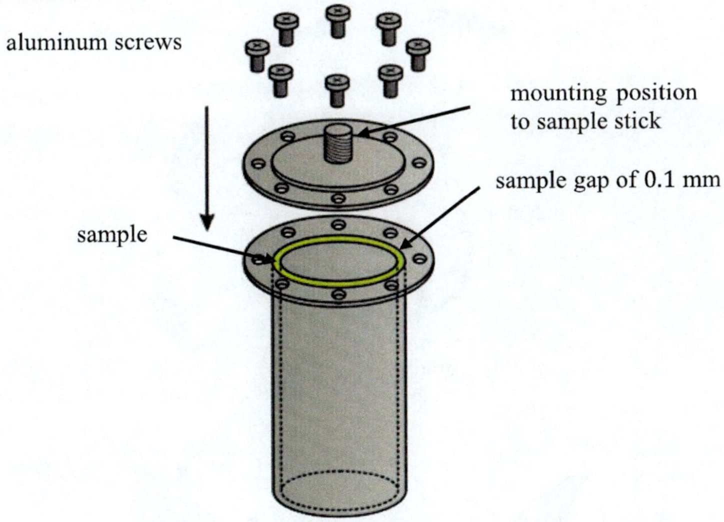

Fig. 3.4 Schematic description of the ANSTO aluminum sample cans, used for all quasi-elastic neutron scattering experiments

transmission. For the data reduction of all three instruments, the program MANTID was used with implemented routines specific for each instrument. Hereby, the data sets are read in and the empty can is subtracted. After that, the data sets are normalized to vanadium, and converted from time-of-flight into energy representation. Finally, the Q-binning is performed and the data sets are corrected for the sample specific resolution. In case of PELICAN, which measures asymmetric spectra, the data sets were additionally mirrored at the y-axis prior to further data treatment.

3.2.1 Neutron Backscattering Instrument EMU

EMU is an indirect geometry neutron backscattering instrument located at ANSTO, Sydney, Australia (Fig. 3.5). Neutrons, coming from the reactor pass the pre-monochromator which selects neutrons of a wavelength, $\lambda = 6.28$ Å. The background chopper separates the neutron beam into several packages, i.e., the neutron beam is pulsed to reduce background scattering. The following graphite chopper deflects the neutrons, in its closed position, to the Doppler drive. The Doppler drive oscillates with a fixed velocity, $v_{\max} = 4.7$ m/s, and shifts the neutrons in energy, creating an energy distribution of $(E_{\mathrm{i}} \pm \delta E)(t)$. The Doppler shifted neutrons pass the graphite chopper in the open position and scatter at the sample into all directions. The small amount of the scattered neutrons, which are fulfilling the Bragg condition after being scattered, is backscattered by the analyzer arrays and pass the sample a second time. The backscattered neutrons are collected at the

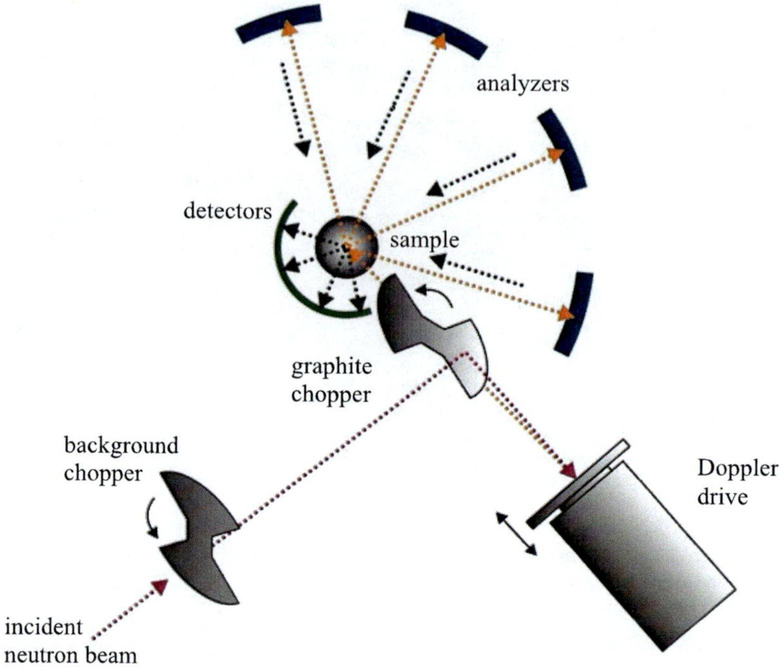

Fig. 3.5 Simplified illustration of the indirect neutron backscattering instrument EMU, including background chopper, graphite chopper, Doppler drive, analyzers, sample, and detectors

detectors. The time dependence of the Doppler drive and therefore of $\delta E(t)$ is known. This allows to obtain the energy transfer ΔE of the sample. These energy transfers at EMU are very small, and therefore, the momentum transfer Q can be approximated by the Eq. (2.14) for elastic scattering.

The elastic fixed window scan gives information about the onset of different relaxation processes occurring in the sample system. For that purpose, typically the Doppler drive is switched off and only scattering with $\Delta E \approx 0$ is detected, while the sample temperature is changing due to the heating or cooling process.

3.2.2 *Time-of-Flight Backscattering Instrument IRIS*

IRIS is an indirect geometry spectrometer located at the pulsed neutron source ISIS, Didcot, United Kingdom. Its simplified illustration is shown in Fig. 3.6. The incident neutron beam from the spallation source is cut into bunches by the chopper, having a frequency of $f = 50$ Hz, resulting in monochromatized and wavelength selected neutron packages with $\lambda = 6.6$ Å. These selected neutrons scatter on the sample into all directions and hit the analyzer arrays. Similar to EMU, those neutrons, fulfilling the Bragg condition are reflected back from the analyzer and collected at the

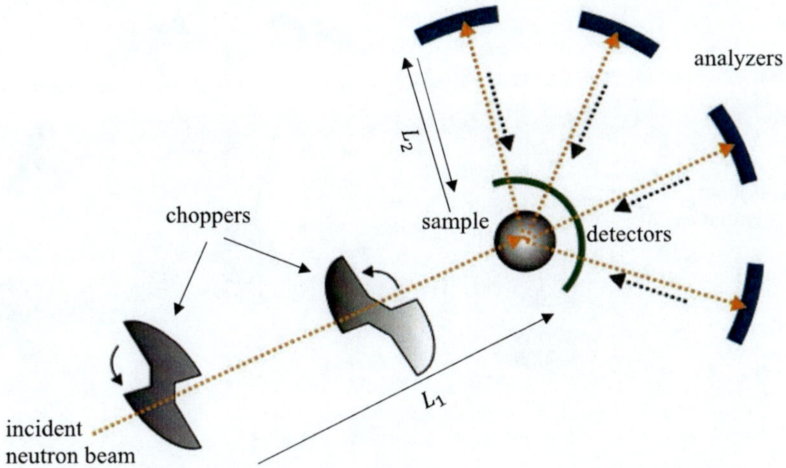

Fig. 3.6 Simplified illustration of the indirect neutron backscattering instrument IRIS, including choppers, analyzers, sample, and detectors, including the distance L_1 and L_2 The real picture of IRIS includes analyzers and detector over a whole circle around the sample

detectors. By knowing the total time-of-flight time, t, of the neutrons together with the distances L_1, source-sample, and L_2, sample-analyzer-detectors, the energy exchange within the sample can be determined.

3.2.3 *Time-of-Flight Spectrometer PELICAN*

Pelican is a direct geometry cold-neutron spectrometer at ANSTO, Sydney, Australia (Fig. 3.7). The incoming neutrons are monochromatized and wavelength selected by a focusing crystal and pulsed by two fermi choppers, which are phased in order to remove $\lambda/2$ from the neutron beam. These pulsed neutrons of $\lambda = 6$ Å scatter at the sample in all directions to be collected at the detectors. The time, t_0, needed for elastic scattering without energy transfer can be determined based on the initial energy, E_i, and the pathlength, which the neutrons need to travel until the detector is reached. The energy transfer can be determined by taking the difference from the "real" flight time, t_{tof}, which was needed including energy exchange and the elastic time, t_0. This results in the scattering intensity depending on scattering angle and time. This instrument is able to detect larger energy transfers compared to EMU and IRIS. Thus, the simple approximation for the momentum transfer, which is based on elastic scattering is not valid. The momentum transfer depends on the incident and scattered wavevector and therefore follows Eq. (2.13).

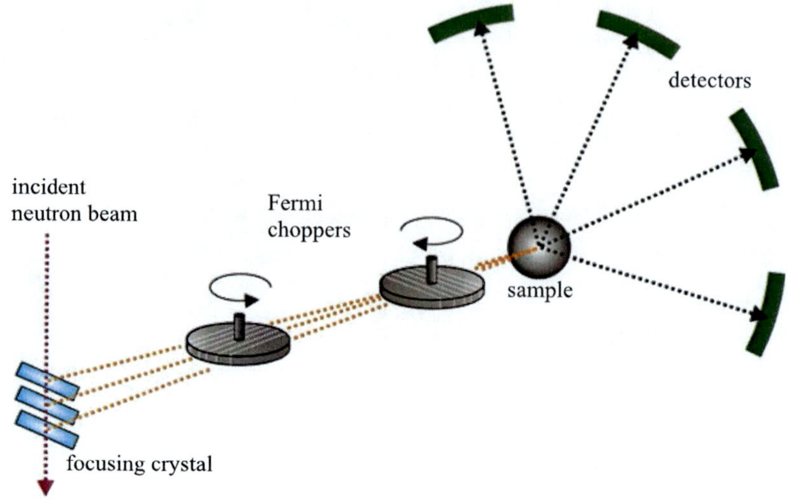

Fig. 3.7 Simplified illustration of the direct time-of-flight spectrometer PELICAN, including focusing crystal, Fermi choppers, sample, and detectors

3.3 Dielectric Spectroscopy

The dielectric spectroscopy measurements were performed with a Broadband Dielectric Alpha Analyzer from Novocontrol GmbH. The instrument operates in a frequency range of 10^{-2}–10^{6} Hz. The available temperature range of $T = -150$ °C to $T = +500$ °C is controlled by a Quatro Cryosystem with a manufacturer specified accuracy of 0.1 °C by using an evaporated liquid nitrogen stream. For the measurements, a temperature range of $T = -140$ °C to $T = -90$ °C with an increment of $\Delta T = 2.5$ °C was enough to move the segmental dynamics through the available frequency range. A waiting time of 3 min prior to each spectrum, ensured a temperature stability equal or better than 0.1 °C. In order to avoid any crystallization, the sample was rapidly cooled down to $T = -140$ °C. Therefore, all the following measurements were performed from low to high temperatures.

For the experiment, the sample is placed between two electrodes, like the dielectric medium for a capacitor and periodic voltage is applied (Fig. 3.8). The sample thickness is controlled by 0.1 mm thick glass spacer.

The periodic voltage can be written as $U^*(\omega) = U_0 \exp(i\omega t)$. This causes a phase shift δ in the complex current $I^*(\omega) = I_0 \exp\left(-i\omega t + \frac{\pi}{2} + \delta\right)$. Both quantities are used to obtain the complex capacity $C_p^* = \frac{I^*(\omega)}{i\omega U^*(\omega)}$. This is connected with the complex dielectric function by

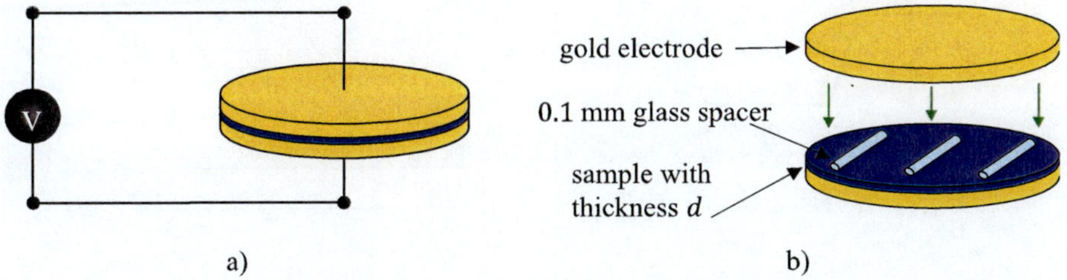

Fig. 3.8 Illustration of (**a**) the measurement principle of dielectric spectroscopy and (**b**) the sample preparation including the glass spacer

$$\epsilon^*(\omega) = \frac{C_p^*}{C_0} \tag{3.1}$$

with $C_0 = \frac{\epsilon_0 A}{d}$ the idle capacity, A the area of the capacitor and d the thickness.

3.4 Fast Field Cycling (FFC) Relaxometry

The fast field cycling relaxometry experiments were performed on the SMARtracer FFC-NMR STELAR, Italy. It is able to measure the longitudinal spin–lattice relaxation time, T_1, dependence on the magnetic relaxation field. The main part of the instrument is a $B = 0.25$ T electromagnet which enables a measurable Larmor frequency range of 10 kHz to 10 MHz by rapidly switching the magnetic detection field. By minimizing external influences, it is possible to extend the frequency range to 5 kHz as the lower limit. For that, the wiring was changed from the original setup, which increases the distance between electric control unit and the electromagnet by 1.5 m. Furthermore, magnetizable parts have been removed as far as possible.

Measurements can be performed in a temperature range from $T = -140$ °C to $T = +140$ °C with accuracy of 0.1 °C (based on information provided by the manufacturer) either controlled by an evaporated liquid nitrogen stream, for temperatures below room temperature, or by a constant compressed air stream for higher temperatures. The temperature range for the samples, studied in this dissertation was $T = -100$ °C to $T = +135$ °C. This is restricted at the low temperature either by the occurring crystallization or by a too short spin–lattice relaxation time, T_1, which than lays within the switching time between relaxation and detection field. The limit for the high temperature was set by the instrument specification. Most of the time, the latter reason causes the restriction, because every sample is quenched in liquid nitrogen prior to the measurements at the low temperature region where crystallization needs to be avoided. For the measurement, the sample is placed in an NMR glass tube with outer diameter of $d = 10$ mm. The filling level is kept between 1.0 and 1.5 cm, to ensure that the sample is placed in the homogenous part of the magnetic field, yielding a good measurement signal.

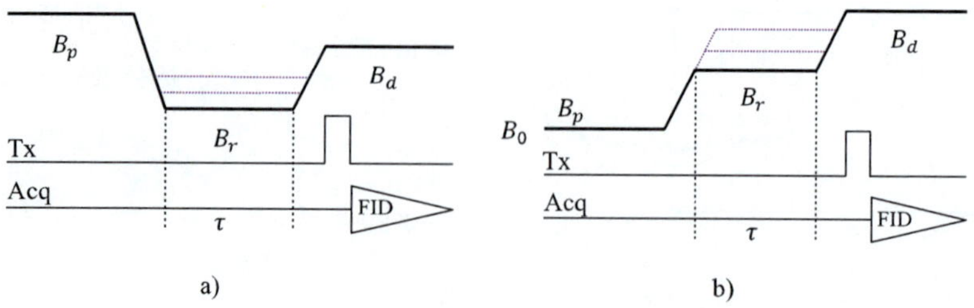

Fig. 3.9 Illustration of the two basic sequences used for fast field cycling measurements, (**a**) pre-polarized (PP) and (**b**) non-polarized (NP) with B_p polarization field, B_r relaxation field, B_d detection field, and delay time, τ. Both of them include the 90° pulse and the free induction decay (FID)

In principle two different measurement sequences need to be used, the so-called Pre-Polarized (PP) and Non-Polarized (NP), to cover the whole frequency range.

In the PP sequence (Fig. 3.9a), the sample is polarized at a high polarization field, B_p, until an equilibrium magnetization is reached. After that, the magnetic field is switched to a lower relaxation field, B_r, and the magnetization will decay until the new equilibrium value is reached. After a delay time τ, a 90° pulse is applied to the sample at a high detection field, B_d, and the free induction decay (FID) is detected. Varying the delay time yields the whole magnetization curve, which usually shows a monoexponentially decay after averaging over a predefined window in the FID signal. Fitting this magnetization curve, results in $R_1 = 1/T_1$ at this specific magnetic field, B_r. Using different B_r the whole relaxation rate depending on the frequency, associated with the relaxation field can be determined.

The NP sequence (Fig. 3.9b) is usually used for higher magnetic fields, whereby the sample equilibrates without an applied external field. After that, the magnetic field is switched to higher relaxation fields, B_r, where the magnetization increases until equilibrium. After the delay time τ the 90° pulse is applied to the sample at the detection field, B_d, and the FID is detected. Similar to PP, by varying the delay time, the whole magnetization curve is measured and described with a monoexponentially increase. This results in the relaxation rate associated with B_r. Changing the relaxation field, the whole relaxation rate over the whole frequency range is measurable.

The selection of NP and PP is based on the step height between polarization field, B_p, and relaxation field, B_r. For good measurement results, this step needs to be as large as possible. At the used setup the change between both sequences occurs at ~3.5 MHz.

References

1. E.P. Gilbert, J.C. Schulz, T.J. Noakes, 'Quokka'—the small-angle neutron scattering instrument at OPAL. Phys. B Condens. Matter **385–386**, 1180–1182 (2006)

2. K. Wood, J.P. Mata, C.J. Garvey, C.-M. Wu, W.A. Hamilton, P. Abbeywick, D. Bartlett, F. Bartsch, P. Baxter, N. Booth, et al., QUOKKA, the pinhole small-angle neutron scattering instrument at the OPAL Research Reactor, Australia: design, performance, operation and scientific highlights. J. Appl. Crystallogr. **51**(2), 294–314 (2018)
3. A. Radulescu, V. Pipich, H. Frielinghaus, M.S. Appavou, KWS-2, the high intensity/wide Q-range small-angle neutron diffractometer for soft-matter and biology at FRM II. J. Phys. Conf. Ser. **351**, 012026 (2012)
4. A. Sokolova, J. Christoforidis, A. Eltobaji, J. Barnes, F. Darmann, A.E. Whitten, L. de Campo, BILBY: time-of-flight small angle scattering instrument. Neutron News **27**(2), 9–13 (2016)
5. A. Sokolova, A.E. Whitten, L. de Campo, J. Christoforidis, A. Eltobaji, J. Barnes, F. Darmann, A. Berry, Performance and characteristics of the BILBY time-of-flight small-angle neutron scattering instrument. J. Appl. Crystallogr. **52**(1), 1–12 (2019)
6. N.R. de Souza, A. Klapproth, G.N. Iles, EMU: high-resolution backscattering spectrometer at ANSTO. Neutron News **27**(2), 20–21 (2016)
7. D. Yu, R. Mole, T. Noakes, S. Kennedy, R. Robinson, Pelican—a time of flight cold neutron polarization analysis spectrometer at OPAL. J. Phys. Soc. Jpn **82**(Suppl A), SA027 (2013)
8. F. Demmel, D. McPhail, C. French, D. Maxwell, S. Harrison, J. Boxall, N. Rhodes, S. Mukhopadhyay, I. Silverwood, V.G. Sakai, F. Fernandez-Alonso, ToF-backscattering spectroscopy at the ISIS facility: status and perspectives. J. Phys. Conf. Ser. **1021**, 012027 (2018)
9. F. Kremer, A. Schönhals, *Broadband Dielectric Spectroscopy* (Springer, Berlin, 2012)
10. R. Kimmich, Principle, purpose and pitfalls of field-cycling NMR relaxometry, in *Field-Cycling NMR Relaxometry*, (Royal Society of Chemistry, London, 2018), pp. 1–41
11. J.S. Higgins, H.C. Benoit, H. Benoît, *Polymers and Neutron Scattering* (Clarendon Press, Oxford, 1996)
12. S.R. Kline, Reduction and analysis of SANS and USANS data using IGOR Pro. J. Appl. Crystallogr. **39**(6), 895–900 (2006)

Chapter 4
Morphology

Abstract The structure of bottlebrush polymers has been investigated in different environments, i.e., in good so9lvent, in good solvent with added linear chains, and in linear host matrices, by using small-angle neutron scattering (SANS). This allowed to identify the shape transition from spherical to elongated objects by increasing the side chain length while keeping the backbone length constant. Once in the spherical region, bottlebrushes can be considered as star polymers with negligible core size and thus, can be seen as shell-only particles if compared to polymer grafted nano particles. In this context, structural changes depending on the molecular weight of the linear host matrices and the bottlebrush volume fraction were investigated. This showed a decreasing radius of the bottlebrushes, induced by the increasing molecular weight of the host matrices. Additionally, clustering starts at the highest volume fraction.

4.1 Data Analysis

4.1.1 Core–Shell Model for Spherical Bottlebrush Polymers

The total scattering intensity, $I(Q)$, is a product of the single particle form factor, $P(Q)$, describing the shape of the object and the structure factor, $S(Q)$, which accounts for the interaction between neighboring particles. For monodisperse systems,

$$I(Q) \propto P(Q) \cdot S(Q) \tag{4.1}$$

Portions of this chapter are reprinted with permission from Bichler, K. J.; Jakobi, B.; Huber, S. O.; Gilbert, E. P.; Schneider, G. J., Structural Analysis of Ultrasoft PDMS-g-PDMS Shell-Only Particles. *Macromolecules* **2020,** *53* (1), 78–89. Copyright (2020) American Chemical Society.

K.J. Bichler, *Morphology and Dynamics of Bottlebrush Polymers*, Springer Theses,
https://doi.org/10.1007/978-3-030-83379-4_4

whereby for dilute systems $S(Q) = 1$ and $I(Q) = P(Q)$ and can be described by a core–shell model originally derived for micelles and star polymers [1–3]. The scattering intensity is

$$I(Q) = \frac{\Phi}{V_{\text{total}}} \cdot P(Q) \tag{4.2}$$

with $P(Q)$ as the single particle form factor of chains attached to a core, following Pedersen [1] with "core" referring to the backbone, "shell" to the side chains, Φ to the volume concentration and V_{total} to the total volume with $V_{\text{total}} = f \cdot V_{\text{shell}} + V_{\text{core}}$.

The single particle form factor, $P(Q)$, can be written as:

$$\begin{aligned} P(Q) &= \Delta\rho_{\text{core}}^2 V_{\text{core}}^2 A(Q)_{\text{core}}^2 \\ &+2f\Delta\rho_{\text{core}}\Delta\rho_{\text{shell}} V_{\text{core}} V_{\text{shell}} A(Q)_{\text{core}} A(Q)_{\text{shell}} \\ &+\Delta\rho_{\text{shell}}^2 V_{\text{shell}}^2 A(Q)_{\text{shell}}^2 f(f-1) \\ &+a \cdot f\Delta\rho_{\text{shell}}^2 V_{\text{shell}}^2 \frac{P(Q)_{\text{blob}}}{1-\nu_{\text{EV}} P(Q)_{\text{blob}}} \end{aligned} \tag{4.3}$$

with $\Delta\rho_{\text{core}} = \rho_{\text{core}} - \rho_{\text{solvent}}$ and $\Delta\rho_{\text{shell}} = \rho_{\text{shell}} - \rho_{\text{solvent}}$ describing the contrast of the core and the shell to the solvent, respectively. The constants ρ_{shell} and ρ_{core} are the scattering length densities of the shell and the core. The volumes V_{core} and V_{shell} are defined as $V_i = \frac{M_{wi}}{\varrho_i N_{\text{A}}}$ with the Avogadro constant $N_{\text{A}} = 6.022 \cdot 10^{23}\ \text{mol}^{-1}$, weight average molecular weight, M_{w_i}, and mass density, ϱ_i, being associated with the core and the shell, f denoting the number of side chains per molecule, and ν_{EV} considers excluded volume effects by taking overlapping chains into account. For polymer melts, excluded volume effects are screened [4].

The form factor of the blob scattering, P_{blob}, can be described with the Beaucage form factor [5] for arbitrary chain statistics, scaled with the prefactor a. It describes the internal chain conformation within the shell, where each side chain is composed of several blobs with increasing size by increasing radial dimension from the center to the outside. In this small region the polymer chain behaves as a Gaussian coil [6]

$$P(Q)_{\text{blob}} = \exp\left(-\frac{Q^2\xi^2}{3}\right) + \left(\frac{d_{\text{f}}}{\xi^{d_{\text{f}}}}\right)\Gamma\left(\frac{d_{\text{f}}}{2}\right)\left(\frac{\text{erf}\left(\frac{Qk\xi}{\sqrt{6}}\right)^3}{Q}\right)^{d_{\text{f}}} \tag{4.4}$$

Here, $k = 1.06$ is an empirical constant, ξ is the radius of gyration, and d_{f} is the fractal dimension of the scattered particle which is typically $d_{\text{f}} = 2$ for melt or Θ-conditions or $d_{\text{f}} = 1.7$ for good solvent condition. This parameter is connected to the Flory exponent, ν, with $d_{\text{f}} = 1/\nu$. The typical value for melt or Θ-condition is $\nu = 0.5$, whereby for good solvent $\nu = 0.588$, as predicted from perturbation theory [7].

The normalized scattering amplitude of the core with a constant density profile can be written as

$$A(Q)_{\text{core}} = \frac{3(\sin(QR_{\text{c}}) - (QR_{\text{c}})\cos(QR_{\text{c}}))}{(QR_{\text{c}})^3} \tag{4.5}$$

with R_{c} as the core radius.

The shell can be described with the normalized scattering amplitude, $A(Q)_{\text{shell}}$, obtained by Fourier transformation of an explicit density profile, $\varphi(r)_{\text{star}}$

$$A(Q)_{\text{shell}} = \frac{\int_{R_{\text{c}}}^{R} 4\pi r^2 \frac{\sin(Qr)}{Qr} \varphi(r)_{\text{star}} \text{d}r}{\int_{R_{\text{c}}}^{R} 4\pi r^2 \varphi(r)_{\text{star}} \text{d}r} \tag{4.6}$$

The shell dimension of the bottlebrushes is finite, therefore the overall radius, R, can be used for the upper integration limit instead of ∞.

The star like density profile, $\varphi(r)_{\text{star}}$, arises from the scaling theory for star polymers [6, 8] where a star consists of three different regions (Fig. 4.1): core, unswollen, and swollen regions.

For Θ-condition or polymer melt with $\nu = 0.5$ the swollen region reduces to $\varphi(r) \propto r^{-1}$ and merges together with the unswollen region. Therefore, only core and shell (unswollen region) are left. For good solvent conditions, all three are present.

The shell can be described by the following ν-dependent density profile, $\varphi(r)_{\text{star}}$ [8–10]:

$$\varphi(r)_{\text{star}} = \frac{r^{\frac{1-3\nu}{\nu}}}{\left(1 + \exp\left(\frac{r-R}{\sigma R}\right)\right)} \tag{4.7}$$

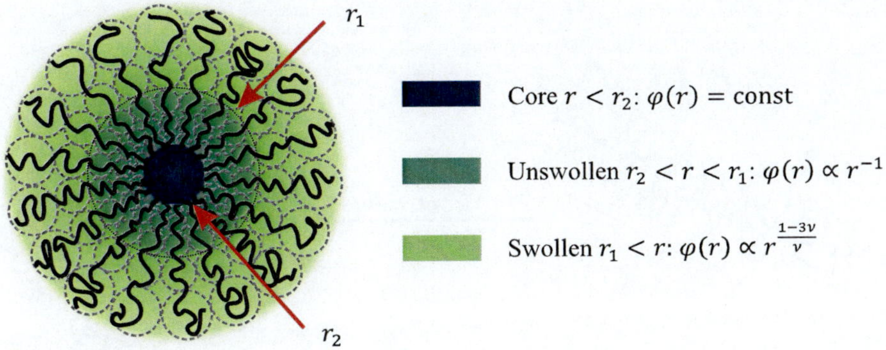

Fig. 4.1 Model for star like polymer representation including blobs of three different density regions originating from scaling theory [8, 11]: Core, unswollen, and swollen regions with ν being the Flory exponent: $\nu = 0.588$ for good solvent and $\nu = 0.5$ for Θ-condition. *(Bichler, K. J., Jakobi, B., Huber, S. O., Gilbert, E. P., Schneider, G. J., Structural Analysis of Ultrasoft PDMS-g-PDMS Shell-Only Particles. Macromolecules 2020, 10.1021/acs.macromol.9b01598)*

Here the radial dependence of the monomer density distribution [11] is multiplied with the Fermi cut-off function to ensure a smooth density decay at R including the smearing parameter, σ, for averaging over all chains.

4.1.2 Form Factor for Elongated Bottlebrush Polymers

For systems having a Kuhn length, ℓ_k, and contour length, L, much larger compared to the cross-sectional radius, R, the form factor can be obtained from the decoupling approximation, whereby the backbone is treated separately from the side chains, i.e., from the cross-section (Fig. 4.2) [12].

The backbone is assumed to have an expanded conformation due to the high grafting density. However, because of the attached side chains, the overall structure is still flexible. Therefore, the backbone can be approximated by a semiflexible polymer chain, whereby the side chains are considered via the cross-section by applying the Fourier transformation of the star like density profile, as written in Eq. (4.7) [13, 14].

The total scattering function for elongated bottlebrush polymer can be written as [13]

$$I(Q) = \Phi V_{\text{total}} \Delta\rho^2 \cdot \left(S_{\text{SC}}(Q) \cdot P_{\text{CS}}(Q) + a \cdot \frac{P_{\text{blob}}(Q)}{1 + \nu_{\text{EV}} P_{\text{blob}}(Q)} \right) \tag{4.8}$$

with volume concentration, Φ, the volume of the whole bottlebrush polymer, V_{total}, and the contrast, $\Delta\rho$. The total volume consists of all side chains together and the backbone. It can be determined by using the molecular weight, M_w.

$$V_{\text{total}} = V_{\text{side chain}} \cdot f + V_{\text{backbone}} \tag{4.9}$$

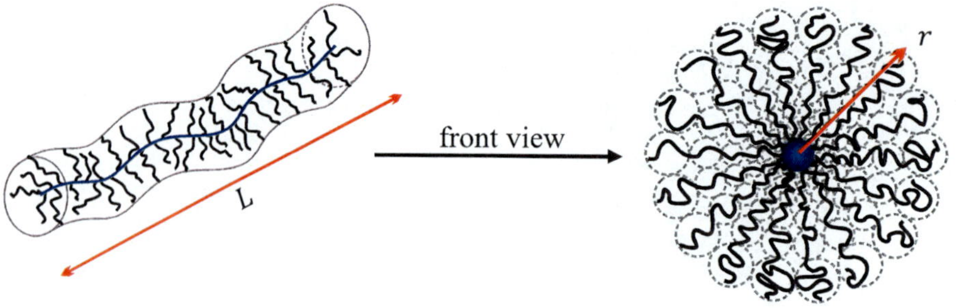

Fig. 4.2 Pictorial representation of an elongated bottlebrush polymer including the front view, representing the cross-section of the molecule. *(Jakobi, B., Bichler, K. J., Sokolova, A., Schneider, G. J., Dynamics of PDMS-g-PDMS Bottlebrush Polymers by Broadband Dielectric Spectroscopy. Macromolecules 2020, 10.1021/acs.macromol.0c01277)*

with the number of side chains, f, and volume, $V_i = \frac{M_{wi}}{\varrho_i N_A}$, associated with side chains and backbone and the respective mass density, ϱ_i. The form factor $S_{SC}(Q)$ stands for an infinite thin, semiflexible polymer chain including excluded volume effects as described by Pedersen et al. [15] in method 3. Additionally, two corrections are included to avoid an unphysical behavior at the low Q-values and to eliminate a kind of a kink in the intermediated Q-region because of swapped terms in the original publication [15]. Both correction are described in Chen et al. [14]. The whole model for the semiflexible polymer chain has three adjustable parameter, the Kuhn length, ℓ_k, the Flory exponent, ν, and the cylinder contour length, L, which is defined as the maximal length which is physical possible, $L = N \cdot b \cdot \sin\left(\frac{\theta}{2}\right)$ [16–18].

The form factor of the cross-section, $P_{CS}(Q)$, can be calculated by a Fourier transformation of the star like density profile, Eq. (4.7). This reads as

$$P_{CS}(Q) = \left| \frac{\int_0^R \varphi(r)_{\text{star}} \cdot J_0(Qr) r \, dr}{\int_0^R \varphi(r)_{\text{star}} r dr} \right|^2 \tag{4.10}$$

with J_0 denoting the zeroth-order Bessel function of first kind and r the radial distance from the backbone. The dimension of our bottlebrush cross-section is finite, therefore the outer radius, R, can be taken for the upper integration limit of the Fourier transformation instead of infinity. For the description of the internal structure within the cross-section, a blob contribution, scaled with the prefactor a, is added to the form factor, including excluded volume considerations. With ν_{EV} overlapping chains are taken into account and get subtracted from the overall contribution. Here the blob form factor, $P_{blob}(Q)$, Eq. (4.4), is identical to the one for the spherical core–shell form factor.

4.2 Results and Discussion

4.2.1 Spherical Bottlebrush Polymer in Good Solvent

The structure of protonated PDMS-g-PDMS bottlebrush polymer was explored under good solvent conditions, in d-toluene, in different volume concentrations, Φ. The main characteristics of the backbone, the side chains, and the whole bottlebrush are summarized in Table 4.1.

The lowest volume concentration, $\Phi = 0.5$ vol%, as displayed in Fig. 4.3a shows a plateau in the low Q-region, associated with the Guinier region followed by a single slope of -4.0 in the intermediate Q-region known as Porod region.

The absence of an additional slope in the transition region from Guinier to Porod suggests a spherical structure of the bottlebrush polymer. In the high Q-region a slope of -1.5 is developed, resembling the structure of the outermost blob [6, 10, 13].

Table 4.1 Weight average molecular weight, M_w, number average molecular weight, M_n, degree of polymerization, DP, and polydispersity, M_w/M_n, of the side chain, the backbone and the whole bottlebrush polymer

Name	M_w (g/mol)	M_n (g/mol)	DP	M_w/M_n
Side chain	8900	8100	110	1.10
Backbone	4230	4160	70	1.02
Bottlebrush	431,000	385,000	70[a]	1.12

[a]Per definition of a bottlebrush polymer

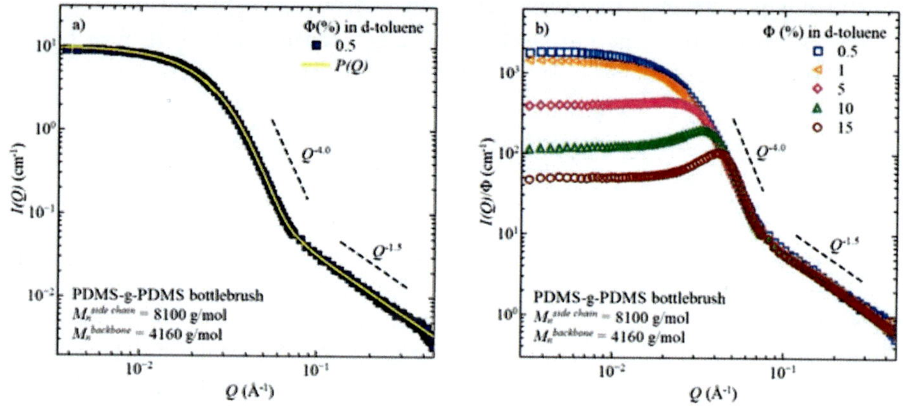

Fig. 4.3 Scattering intensities, *I(Q)*, vs. momentum transfer, *Q*, of PDMS-g-PDMS bottlebrush polymers under good solvent condition. (**a**) Lowest volume concentration representing the pure form factor, *P(Q)*, described with the spherical core–shell model under the assumption of negligible core (full yellow line). (**b**) Concentration series normalized to the respective volume concentration Φ

Based on scaling theory for good solvent condition, the blob size increases with increasing radius form the center to the outermost part [6]. Furthermore, the slope is connected with the Flory parameter by $d_f = \frac{1}{\nu}$. Here, $d_f = 1.50 \pm 0.02$ which gives $\nu = 0.\overline{6}$. This value is close to the predicted value of $\nu = 0.588$ [7, 19, 20]. For this low concentration it is assumed that no contribution of a structure factor is present and therefore it can be described with the pure form factor, *P(Q)*, only. The form factor used for the description is based on Eq. (4.3) with the star like density profile (Eq. 4.7) and the blob contribution for the high *Q*-region, as described in Sect. 4.1.1.

For the modeling, the following approximation is used. The core size as well as the molecular weight of the backbone is assumed to be very small compared to the overall bottlebrush and therefore it can be neglected and is not included in the SANS model function. As seen in Fig. 4.3a, with this approximation the data can be described very well and the resulting fit parameters are summarized in Table 4.2.

The number of side chains from the SANS model agrees well with the number obtained from Gel Permeation Chromatography (GPC) measurements, $f_{GPC} = 47$, which can simply be calculated from the number average molecular weight, M_n, of

Table 4.2 Best fit values for PDMS-g-PDMS bottlebrush polymers in d-toluene, obtained by modeling the single particle form factor at the lowest measured concentration $\Phi = 0.5\%$. All standard deviations are smaller than 1%

Parameter	PDMS-g-PDMS bottlebrush
Overall radius R (Å)	72.5
Radius of outermost blob ξ (Å)	8.8
Scaling parameter of blob a	0.06
Number of side chains f	47
Excluded volume parameter ν_{EV}	−0.7

the bottlebrush knowing the respective backbone and side chains. Considering the radius of gyration of the pure side chains of $R_g \approx 33$ Å, estimated from a similar polymer in solution [21], the side chains stretch by a factor of around 1.1, to obtain the full dimension as determined by the model function.

With increasing concentration, the scattering intensity evolves a correlation peak in the Porod region as seen in Fig. 4.3b. Starting at a volume concentration of $\Phi = 5\%$ the correlation peak intensifies up to the highest measured concentration, $\Phi = 15\%$. This peak arises due to interparticle correlations and can be described with the assumption of an ultrasoft potential, proposed by Likos et al. [22] which reads as

$$\beta V_{\mathrm{BB}}(r) = \left(\frac{5}{18}\right) f^{\frac{3}{2}} \begin{cases} \left(\dfrac{1}{1+\dfrac{\sqrt{f}}{2}} - \ln\left(\dfrac{r}{\sigma_{\mathrm{B}}}\right)\right) & \text{for } r \le \sigma_{\mathrm{B}} \\ \dfrac{1}{1+\dfrac{\sqrt{f}}{2}}\left(\dfrac{\sigma_{\mathrm{B}}}{r}\right) \exp\left(-\dfrac{\sqrt{f}(r-\sigma_{\mathrm{B}})}{2\sigma_{\mathrm{B}}}\right) & \text{for } r > \sigma_{\mathrm{B}} \end{cases} \tag{4.11}$$

with f the number of side chains, σ_{B} the so-called corona diameter, equivalent to the bottlebrush extension, r the distance between two bottlebrush centers, and $\beta = (k_{\mathrm{B}}T)^{-1}$ where k_{B} is the Boltzmann constant and T is the temperature.

The scattering curve can be described with a combined model, consisting of the form factor, $P(Q)$, obtained from the lowest concentration and the structure factor, $S(Q)$, as displayed in Fig. 4.4a. The latter one is obtained by solving the Ornstein–Zernike equation with the ultrasoft potential and the Roger-Young closure relation [23] as reported in the literature [22].

Deviations are visible in the mid Q-region. These could originate in the missing resolution contribution, which is accounted for in the form factor, but not in the theoretical description of the structure factor. The peak position of the pure structure factor, obtained by division of the high concentration spectra by the form factor, resembles the center of mass distance of two neighboring bottlebrushes with $d = \frac{2\pi}{Q_{\max}}$. By looking at the dependence of the distance, d, related to the volume concentration, Φ, a power law evolves (Fig. 4.4b).

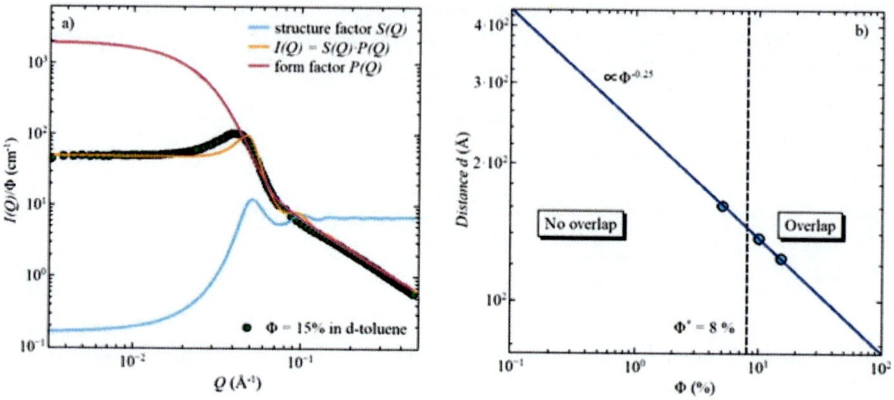

Fig. 4.4 (**a**) Scattering intensities, $I(Q)$, vs. momentum transfer, Q, of PDMS-g-PDMS bottlebrush polymers of the highest volume concentration $\Phi = 15\%$ described with form factor $P(Q)$ and structure factor $S(Q)$. (**b**) Distance, d, vs. volume concentration, Φ, for the PDMS-g-PDMS bottlebrush polymers including the overlap concentration Φ^*

A similar behavior was found by Willner et al. [24] for 8 and 18 armed polyisoprene star polymers in solution. Hereby, a power law of $d \propto \Phi^{-0.25}$ describes the concentration dependent distance d. This power law can also be applied to the experimental data obtained for the PDMS-g-PDMS bottlebrush in solution, even if there are only three data points available (Fig. 4.4b). However, for this particular behavior, a theoretical explanation is missing [24]. The reason could be the nature of the ultrasoft potential. A simulation of several structure factors based on Eq. (4.11) for different concentrations results in the same dependency of $d \propto \Phi^{-0.25}$. Furthermore, an experimentally estimation of the overlap concentration, Φ^*, is possible by taking the diameter, $D = 2R$, as the closest distance without overlap. Considering the power law dependence, $d \propto \Phi^{-0.25}$, the overlap concentration ends up at $\Phi^* = 8.0\%$. For all concentrations with $\Phi > 8.0\%$, the side chains will overlap which results in shorter distances between two neighbored bottlebrushes. This allows to estimate the highest possible interpenetration, by assuming a validity of the power law up to $\Phi = 100\%$. The closest possible distance for this system is $D_{\text{min}} = 77.1$ Å which is equivalent to an overlap of ~94%.

In the next step linear deuterated PDMS chains of $M_n = 3600$ g/mol were added to the solution of PDMS-g-PDMS bottlebrush in a d/h-toluene mixture to follow the transition to the linear melt environment without solvent. The ratio for the d/h-toluene was chosen so that only the protonated bottlebrush polymer was visible and the scattering length density of deuterated linear PDMS, $\rho = 4.958 \cdot 10^{10}$ 1/cm^2, was contrast matched. This ratio was determined by a separate contrast variation experiment, whereby the content of deuterated and protonated toluene was systematically changed in order to determine the zero contrast condition. Herefore, the intensity difference, ΔI, of two selected Q-values, Q_1 and Q_2, was taken and analyzed regarding their volume concentration of protonated toluene, Φ_H (Fig. 4.5).

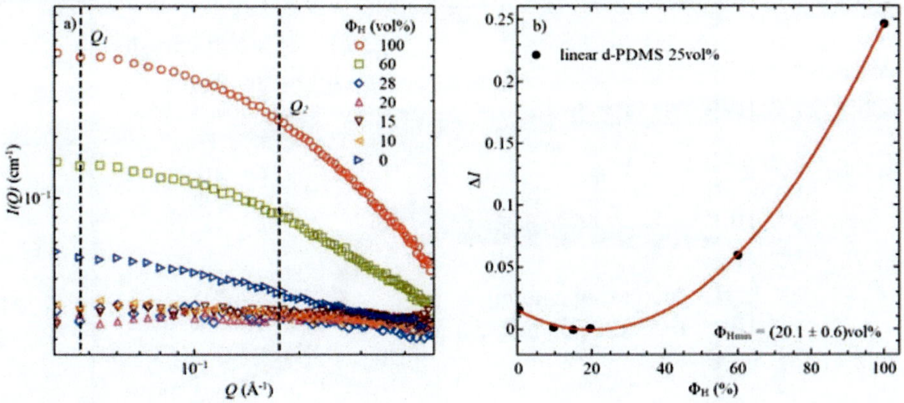

Fig. 4.5 Contrast variation experiment for deuterated linear PDMS in d/h-toluene. (**a**) Scattering intensity, $I(Q)$, vs. momentum transfer, Q, of linear deuterated PDMS in different volume concentrations Φ_H of protonated toluene with $\Phi_H + \Phi_D = 100\%$ including the two selected Q-values for determining the contrast parabola. (**b**) Difference in scattering intensity, $\Delta I = I(Q_1) - I(Q_2)$, vs. volume concentration of protonated toluene, Φ_H. The full red line represents the model function for determining the correct matching condition of $\Phi_H = (20.1 \pm 0.6)$vol%. *(Bichler, K. J., Jakobi, B., Huber, S. O., Gilbert, E. P., Schneider, G. J., Structural Analysis of Ultrasoft PDMS-g-PDMS Shell-Only Particles. Macromolecules 2020, 10.1021/acs.macromol.9b01598)*

By using the following model function

$$\Delta I = a(\Phi_H - \Phi_{H_{min}})^2 + c \tag{4.12}$$

whereby a and c are fit constants, the value for the appropriate volume concentration of protonated toluene, determined as the minimum value of the quadratic relationship, represented as ΔI vs. Φ_H in Fig. 4.5b can be obtained experimentally. The resulting matching condition was $\Phi_{H_{min}}^{exp} = (20.1 \pm 0.6)$ vol%. This differs from the theoretical calculation of the matching point of $\Phi_{H_{min}}^{calc} = 14.9$vol%, based on the elemental composition of PDMS. The different value may arise from the low molecular weight of the PDMS. Here end group effects contribute to the scattering length density, which are not accounted for in the calculation. Furthermore, considering Fig. 4.5a, a broad range of zero contrast with $\Phi_H = 10 - 28$vol% is achievable, as indicated by the overlapping scattering curves.

A constant volume fraction of bottlebrush polymer, $\Phi_{BB} = 12$vol%, was mixed with different concentrations of deuterated linear PDMS chains under matching conditions of the deuterated chains. This ensures that only scattering of the bottlebrush polymer is observed.

As seen in Fig. 4.6, with an increasing amount of linear polymer chains, the evolved structure peak in the Porod region starts to weaken and the forward scattering shifts to higher intensities. The high Q-region is not influenced by this effect and all scattering intensities overlap in the region with the same power law of

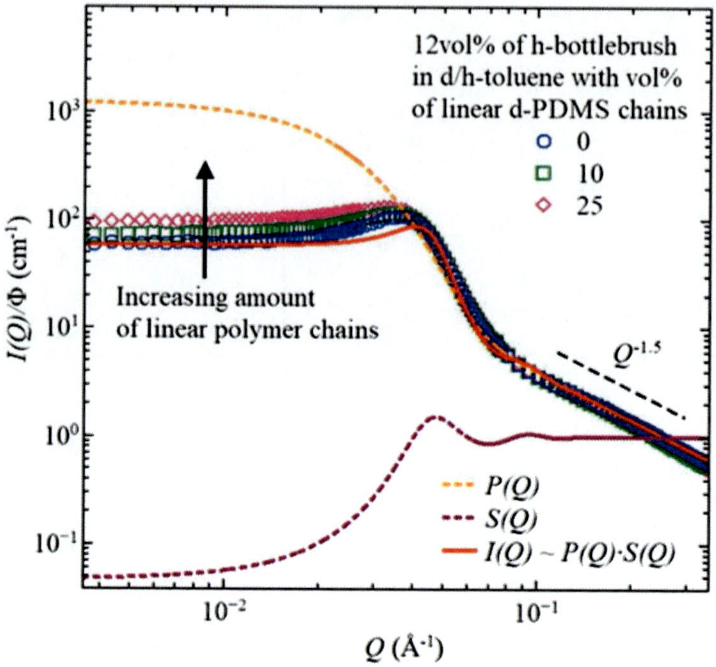

Fig. 4.6 Scattering intensity, $I(Q)$, vs. momentum transfer, Q, for a constant volume fraction of PDMS-g-PDMS bottlebrush polymer in d/h-toluene with added linear d-PDMS chains. *(Bichler, K. J., Jakobi, B., Huber, S. O., Gilbert, E. P., Schneider, G. J., Structural Analysis of Ultrasoft PDMS-g-PDMS Shell-Only Particles. Macromolecules 2020, 10.1021/acs.macromol.9b01598)*

$Q^{-1.5}$ which was found earlier for the sample dissolved in pure d-toluene. This emphasizes that only the correlation between neighbored bottlebrushes is influenced by the addition of linear chains, but not the internal structure within the shell. The linear chains occupy space between side chains and reduce the internal fluctuation of the side chains from overlapping bottlebrush polymers. This causes the decrease of the correlation peak [25].

The scattering intensity without added linear chains can be described in the same way as in d-toluene, the only difference is the changed contrast, due to the d/h-toluene. For the form factor, the parameters obtained from the previous experiment were used, only the scattering length densities, ρ, was changed to consider the d/h-toluene mixture.

The description of the other scattering intensities with added linear chains is too complicated to perform, because of two additional interaction potentials, describing the interaction between two linear chains and the interaction between bottlebrush polymer and linear chains which needs to be taken into account.

The reduction of the correlation peak is even more pronounced if the whole solvent is substituted by linear polymer chains. In this case the correlation peak is completely gone.

4.2.2 Spherical Bottlebrush Polymer in Different Molecular Weight Linear Host Matrices

In this step the whole solvent was substituted by linear PDMS chains, whereby the bottlebrush polymer was deuterated and the linear host matrices protonated. This changes the quality of the solvent into a Θ- or melt condition with the related Flory exponent of $\nu = 0.5$.

The synthesis of the used samples in this chapter was part of the dissertation of Bruno Jakobi [26] and will be not included here, only the important information, helpful for the understanding of this chapter will be included. For the more detailed synthesis description, I would like to refer either to the dissertation of Bruno Jakobi [26] or to the related publication [27].

In this experiment, three different molecular weight linear host matrices were used, according to Table 4.3, giving information about the changes in size and miscibility. This mixture of architectures effectively represents shell-only particles immersed in linear polymer melts. It allows to study the molecular weight dependent changes on the morphology of only the shell, without any influences arising from core contributions, i.e., shell-only particles.

The used bottlebrush polymer is similar to the previous one, but the side chains are deuterated with protonated end groups, i.e., (h-PDMS)-g-(d-PDMS) bottlebrush polymer (backbone-g-side chain). In order to compare the structural behavior to the previous experiments (bottlebrush polymers in good solvent) the molecular weight for side chains as well as backbone was chosen to be similar to the previous case as summarized in Table 4.4.

The grafting reaction towards the bottlebrush polymer resulted in a grafting density of 56% which is well above the threshold from comb to bottlebrush polymer.

The used form factor for the data modeling is equivalent to the previously used one. However, considering the scattering lengths of each part of our (h-PDMS)-g-(d-PDMS) bottlebrush polymer with deuterated side chains, having protonated end groups at the dangling ends, there are three different contrasts (Fig. 4.7a): core, shell, and end group. By including every contrast into the form factor, $P(Q)$ has a core-shell-shell structure, where the second shell consists only of the end group

Table 4.3 Weight average molecular weight, M_w, number average molecular weight, M_n, polydispersity, M_w/M_n, and the calculated radius of gyration, R_g^{calc}, of the linear poly(dimethylsiloxane) (PDMS) polymer melt, used as the host matrix for the PDMS-g-PDMS bottlebrush polymer *(Bichler, K. J., Jakobi, B., Huber, S. O., Gilbert, E. P., Schneider, G. J., Structural Analysis of Ultrasoft PDMS-g-PDMS Shell-Only Particles. Macromolecules 2020, 10.1021/acs.macromol.9b01598)*

Name	M_w (g/mol)	M_n (g/mol)	M_w/M_n	R_g^{calc} (Å)[a]
lin-h-PDMS9k	10,600	8700	1.21	30.2
lin-h-PDMS38k	44,700	38,000	1.17	62.2
lin-h-PDMS68k	69,400	68,000	1.02	77.5

[a]Based on the statistical segment length of PDMS, $b = 6.2$ Å [21]

Table 4.4 Weight average molecular weight, M_w, number average molecular weight, M_n, degree of polymerization, DP, and polydispersity, M_w/M_n, of the side chain (d-PDMS), the backbone (h-PDMS) and the (h-PDMS)-g-(d-PDMS) bottlebrush polymer *(Bichler, K. J., Jakobi, B., Huber, S. O., Gilbert, E. P., Schneider, G. J., Structural Analysis of Ultrasoft PDMS-g-PDMS Shell-Only Particles. Macromolecules 2020, 10.1021/acs.macromol.9b01598)*

Name	M_w (g/mol)	M_n (g/mol)	DP	M_w/M_n
Side chain	11,400	10,800	136	1.07
Backbone	4230	4160	70	1.02
Bottlebrush	446,000	404,000	70[a]	1.10

[a]Per definition of a bottlebrush polymer

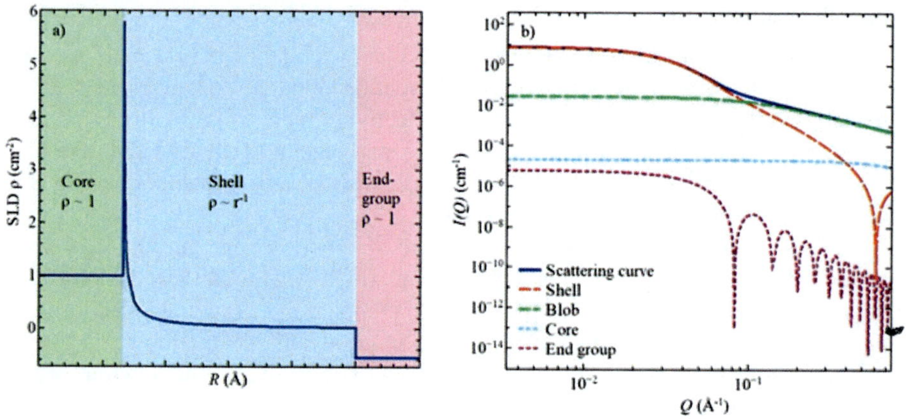

Fig. 4.7 (**a**) Scattering length density, ρ, vs. radius, R, of the protonated core, the deuterated shell, and the protonated end group. Core and end groups have a constant density profile, whereas the shell has a radial dependence based on scaling theory (Eq. (4.7) for Θ-condition). (**b**) Scattering intensity, $I(Q)$, vs. momentum transfer, Q, of the model function together with the intensities from the individual contributions. *(Bichler, K. J., Jakobi, B., Huber, S. O., Gilbert, E. P., Schneider, G. J., Structural Analysis of Ultrasoft PDMS-g-PDMS Shell-Only Particles. Macromolecules 2020, 10.1021/acs.macromol.9b01598)*

contribution. However, the end group still belongs to the side chains which build up the shell of the bottlebrush polymer, just because of the different contrasts it is named as a second "shell." The scattering intensities of the core and the end group shell are at least one order of magnitude weaker in comparison to that of the shell (Fig. 4.7b).

Therefore, the (h-PDMS)-g-(d-PDMS) bottlebrush can be approximated by only considering the shell and the blob contribution. As the later analysis demonstrates, these two parts are sufficient to describe the whole scattering curve. Accordingly, the single particle form factor, $P(Q)$, can be written as

$$P(Q)_{BB} = \Delta\rho^2_{shell} V^2_{shell} A(Q)^2_{shell} f(f-1) + a \cdot f \Delta\rho^2_{shell} V^2_{shell} P(Q)_{blob} \quad (4.13)$$

with the same expressions for the scattering amplitude as described earlier. Here the density profile of the shell can be described by

$$\varphi(r)_{\text{star}} = \frac{r^{-1}}{\left(1 + \exp\left(\frac{r-R}{\sigma R}\right)\right)} \tag{4.14}$$

suitable for Θ- or melt conditions with the related Flory exponent of $\nu = 0.5$.

For systems composed of different architectures of considerable size and different isotope labelings—here deuterated bottlebrush polymers and protonated linear chains—the random phase approximation (RPA) allows to weight all contributions including the interactions between them

$$\frac{(\Delta\rho)^2}{I(Q)} = \frac{1}{\Phi_{\text{BB}} V_{\text{BB}} \widetilde{P}(Q)_{\text{BB}}} + \frac{1}{\Phi_{\text{LC}} V_{\text{LC}} P(Q)_{\text{Debye}}} - 2\chi/V_0 \tag{4.15}$$

Here Φ_i and V_i correspond to the respective volume fraction and molecular volume of the bottlebrush polymer with $V_{\text{BB}} = f \cdot V_{\text{shell}}$ and the linear chains, V_0 is the geometric average monomer volume with $V_0 = \sqrt{V_{\text{H}} V_{\text{D}}}$ where V_{H} and V_{D} are the protonated and deuterated monomer volume, respectively [28]. $P(Q)_{\text{Debye}}$ is the form factor of the linear chains of the host matrix and is given by the Debye formula

$$P(Q)_{\text{Debye}} = \frac{2}{\left(Q^2 R_{\text{g}}^2\right)^2} \left(Q^2 R_{\text{g}}^2 - 1 + \exp\left(-Q^2 R_{\text{g}}^2\right)\right) \tag{4.16}$$

with the radius of gyration R_{g}. R_{g} was calculated based on the relationship $\left\langle R_{\text{g}}^2 \right\rangle = \frac{1}{6} \left\langle R_{\text{ee}}^2 \right\rangle$ and the statistical segment length $b = 6.2$ Å of PDMS [21]. For all modeling, R_{g} was fixed to theoretical values. The form factor, used for considering the bottlebrush contribution changes from Eq. (4.13) to

$$\widetilde{P}(Q)_{\text{BB}} = A(Q)^2_{\text{shell}} \frac{1}{f} (f - 1) + a \cdot \frac{1}{f} P(Q)_{\text{blob}} \tag{4.17}$$

Most of the parameters presented above are calculated from the chemistry of the molecules. As seen in Fig. 4.7a, the whole scattering curve can be described with the shell and the blob contribution only. Each part can be described with a set of two parameters independent of each other.

The scattering intensities, divided by their respective bottlebrush volume fraction are shown for all three different molecular weight host matrices, in Figs. 4.8 and 4.9.

The scattering intensities of (h-PDMS)-*g*-(d-PDMS) bottlebrushes in lin-h-PDMS9k (Fig. 4.8a) superimpose in the intermediate to high Q-region, but show substantial differences at low Q. An extrapolation of the concentration series to $\Phi \rightarrow 0$ (yellow stars), by using $\Phi = 0.5\ \%$, $1\ \%$, and 5%, indicates that there is a

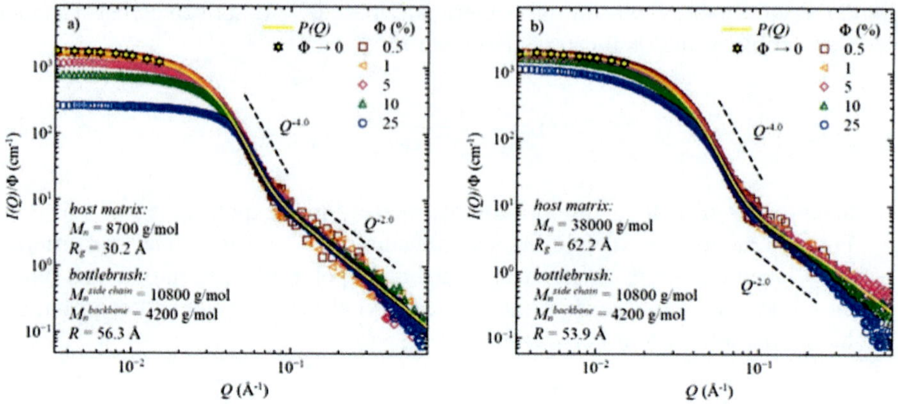

Fig. 4.8 Scattering intensities, $I(Q)$, vs. momentum transfer, Q, of PDMS-g-PDMS bottlebrush polymers in (**a**) lin-h-PDMS9k and (**b**) lin-hPDMS38k matrices in double-logarithmic representation. Concentration series normalized to the respective volume fraction, Φ, including the extrapolated values that represent $I(\Phi = 0\%)$. The full yellow line represents a core–shell model used to describe the lowest scattering intensity, $\Phi = 0.5\%$, based on Eq. (4.15). The errors are within the symbol size and omitted for clarity. *(Bichler, K. J., Jakobi, B., Huber, S. O., Gilbert, E. P., Schneider, G. J., Structural Analysis of Ultrasoft PDMS-g-PDMS Shell-Only Particles. Macromolecules 2020, 10.1021/acs.macromol.9b01598)*

negligible or no structure factor, $S(Q)$, contribution at the lowest measured volume fraction [29].

This region of small momentum transfer is particularly sensitive to the structure factor, which would be visible as reduced forward scattering through a repulsive interaction. The scattering at the lowest concentration of $\Phi = 0.5\%$ is therefore consistent with the dilute regime enabling access to the particle form factor, $P(Q)$.

The low Q-range corresponds to the Guinier region, whereas the intermediate Q^{-4} behavior reflects the Porod law. As seen in the scattering curves, the transition between Porod and Guinier region is not accompanied by an additional region of a different slope. This suggests that the structure of the PDMS based bottlebrushes is consistent with spherical objects. The Q^{-2} decay in the high Q-region is a signature of blob scattering [6], whereby only the outermost blob is visible in SANS experiments [13]. One may therefore pose the question as to whether the size of the blobs changes with distance from the center. In order to tackle this question, it needs to be recalled that the lin-h-PDMS9k matrix and the side chains of the bottlebrush have the same molecular weight. Therefore, it can be assumed that the mixture physically acts as a concentrated star solution, by considering the bottlebrushes and the linear chains together. This statement is based on the classification of Daoud et al. [6] which considers two regions—dilute and concentrated solutions—for scaling theory. These assumptions will be used further. Based on the scaling laws for this region (concentrated region), all blobs have the same size once they start to overlap [6, 8, 30]. Additionally, the power law in the blob region observed in our experiments follows

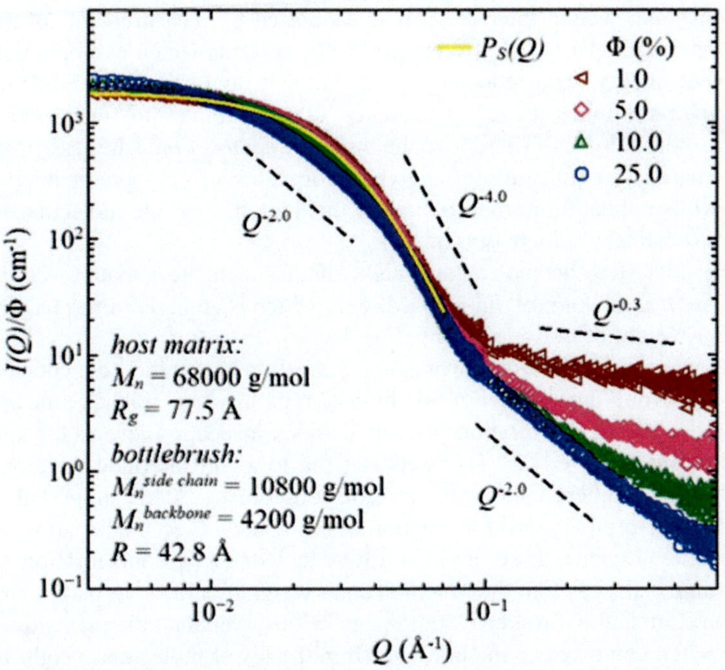

Fig. 4.9 Scattering intensities, $I(Q)$, vs. momentum transfer, Q, of PDMS-g-PDMS bottlebrush polymers in lin-h-PDMS68k matrix in a double-logarithmic representation. Concentration series normalized to the respective volume fraction, Φ. The full yellow line represents the description of the form factor for $\Phi = 1.0\%$ by only taking the shell contribution of the core−shell model (Eq. 4.15). *(Bichler, K. J., Jakobi, B., Huber, S. O., Gilbert, E. P., Schneider, G. J., Structural Analysis of Ultrasoft PDMS-g-PDMS Shell-Only Particles. Macromolecules 2020, 10.1021/acs. macromol.9b01598)*

$Q^{-d_f} = Q^{-2}$. It is connected to the Flory exponent ν by $d_f = 1/\nu$. Therefore, the experimental $\nu = 0.5$ indicates a Θ-condition for all concentrations.

As illustrated in Fig. 4.8b, increasing the molecular weight of the linear host matrix by a factor ~4 (M_n = 38 kg/mol, lin-h-PDMS38k) does not substantially affect the Q dependence of $P(Q)$, i.e., the bottlebrush appears to stay spherical. At least at higher bottlebrush volume fractions, the spherical structure with a smooth surface ($I \propto Q^{-4}$) is still maintained. A more detailed comparison with Fig. 4.8a reveals that the concentration dependence changes, since the reduction of the forward scattering with increasing the concentration is less pronounced than observed in the short matrix (lin-h-PDMS9k). In the high Q-region, the power law changes with increasing concentration, but still reaches $d_f = 2$ at $\Phi = 25\%$.

The change of the slope indicates a volume fraction-dependent solvent quality. In the lin-h-PDMS38k matrix the Θ-condition with $\nu = 1/2$ is only reached for the highest volume fraction, represented by a power law of Q^{-2}. This seems to be plausible because the mutual interpenetration of the side chains increases with increasing bottlebrush concentration. This leads to the fact that more chains of equal molecular weight interact. This is supported by measurements of the bottlebrushes in matrix chains of different molecular weights which exhibit a decrease of the solvent quality. The reduction of the solvent quality in the higher molecular weight matrix implies a greater tendency of the samples to form aggregates or different morphologies. Therefore, the structure factor would be less pronounced, though, there is no indication of aggregate formation or non-spherical structures in the experimental data. In order to explore this idea further, the molecular weight of the matrix chains was increased further.

In the next step the molecular weight of the linear host matrix was increased further, to $M_n = 68$ kg/mol, lin-h-PDMS68k, which is around 7 times larger than the molecular weight of the side chains (Fig. 4.9).

Most apparent is the almost vanishingly small concentration dependence at low Q, and the strong dependence in the high Q-region which reaches a power law of Q^{-2} at the highest concentration. A more detailed inspection reveals the appearance of an additional power law Q^{-2} between the low and intermediate Q-region for $\Phi = 25\%$, the highest volume fraction of bottlebrushes. This suggest the starting point of clustering [31] or the formation of elongated objects, consisting of several bottlebrushes. Despite these obvious differences, the low concentration sample is very similar to the cases in the low molecular weight matrices. In particular, a direct transition from Guinier to Porod region can be observed, and the intensities are very similar, when compared to the bottlebrush in the lower molecular weight matrices.

In this case the data modeling was done with $\Phi = 1.0\%$ (due to leaking sample of the $\Phi = 0.5\%$ during transport). Based on the behavior of the bottlebrush polymer in lin-h-PDMS9k and lin-h-PDMS38k, it can be assumed that there is no contribution of the structure factor even at $\Phi = 1.0\%$ and the assumption of a pure form factor contribution is still satisfied, at least within a good approximation.

The best model description of the experimental data can be achieved by a modified core shell model with a core of negligible size, used together with the random phase approximation (RPA) (Eq. 4.15). The analysis supports the argument accompanying Figs. 4.8 and 4.9 that the samples with a low bottlebrush concentration can be well described by the particle form factor, $P(Q)_{\mathrm{bb}}$, only. This form factor only contains the shell and the blob contributions (Eq. 4.13).

As expanded below, the density profile depends on the bottlebrush concentration and the host matrix molecular weight. Therefore, the average radius, R_{avg}, can be calculated from the overall radius, R, obtained by using the model function and the smearing parameter, σ, within the density profile, $\varphi(r)_{\mathrm{star}}$ (Eq. 4.14) [32]

$$R_{\mathrm{avg}} := \frac{\int_0^R r^3 \cdot \varphi(r)_{\mathrm{star}} \mathrm{d}r}{\int_0^R r^2 \cdot \varphi(r)_{\mathrm{star}} \mathrm{d}r} \tag{4.18}$$

The best fit values for $\Phi = 0.5\%$ of lin-h-PDMS 9k and in lin-h-PDMS38k are summarized in Table 4.5 and for $\Phi = 1.0\%$ in Table 4.6. As Fig. 4.10 illustrates, R_{avg} decreases linearly with increasing molecular weight of the host matrix. The average radius for the highest molecular weight is already identical with the anticipated value

Table 4.5 Best fit values of the bottlebrush polymers, mixed with lin-h-PDMS9k and lin-h-PDMS38k, obtained by modeling the single particle form factor at the lowest concentration measured $\Phi = 0.5\%$. All standard deviations are smaller than 1% *(Bichler, K. J., Jakobi, B., Huber, S. O., Gilbert, E. P., Schneider, G. J., Structural Analysis of Ultrasoft PDMS-g-PDMS Shell-Only Particles. Macromolecules 2020, 10.1021/acs.macromol.9b01598)*

Parameter	lin-h-PDMS9k ($\Phi = 0.5\%$)	lin-h-PDMS37k ($\Phi = 0.5\%$)
Overall radius R (Å)	56.3	53.9
Radius of outermost blob ξ (Å)	14.2	8.7
Scaling parameter of blob a	0.14	0.08
Smearing parameter σ	0.36	0.39
Number of side chains f	39.0	40.0
Average radius R_{avg} (Å)	107.5	97.9

Table 4.6 Best fit values of all systems obtained by modeling the single particle form factor at $\Phi = 1.0\%$ (cf. Appendix A, Fig. A.1). For the largest host matrix, the blob size was not possible to determine due to an unphysical slope in the high Q. All standard deviations are smaller than 1% *(Bichler, K. J., Jakobi, B., Huber, S. O., Gilbert, E. P., Schneider, G. J., Structural Analysis of Ultrasoft PDMS-g-PDMS Shell-Only Particles. Macromolecules 2020, 10.1021/acs.macromol.9b01598)*

Parameter	lin-h-PDMS9k ($\Phi = 1.0\%$)	lin-h-PDMS37k ($\Phi = 1.0\%$)	lin-h-PDMS68k ($\Phi = 1.0\%$)
Overall radius R (Å)	56.3	53.1	42.8
Radius of outermost blob ξ (Å)	14.2	8.7	–
Scaling parameter of blob a	0.14	0.09	–
Smearing parameter σ	0.36	0.39	0.53
Number of side chains f	39.0	40	40.0
Average radius R_{avg} (Å)	107.5	85.6	55.6

for the lowest limit radius, $R_{\text{avg}}^{\text{low}}$. To estimate $R_{\text{avg}}^{\text{low}}$ it can be assumed that the minimum value of the mass density cannot be below the physical density, $\varrho_{\text{d - PDMS}} = 1.02$ g/mol, of linear deuterated PDMS. From $\varrho_{\text{d-PDMS}} = \frac{M}{V}$ with $M = M_{\text{w}}/N_{\text{A}}$, replacing V by the volume of a sphere, $V = \frac{4\pi}{3}\left(R_{\text{avg}}^{\text{low}}\right)^3$, yields:

$$R_{\text{avg}}^{\text{low}} = \left(\frac{3}{4\pi N_{\text{A}}} \frac{M_{\text{w}}^{\text{bottlebrush}}}{\varrho_{\text{d-PDMS}}}\right)^{\frac{1}{3}} \tag{4.19}$$

By using Eq. (4.19) and taking $M_{\text{w}}^{\text{bottlebrush}}$ from Table 4.4, a value of $R_{\text{avg}}^{\text{low}} = 55.7$ Å can be found. Based on the calculated value for the lower limit of the radius, $R_{\text{avg}}^{\text{low}}$, the size obtained for the bottlebrush polymer in the highest molecular weight host matrix, lin-h-PDMS68k, is the smallest possible value for our system. This statement has a very important consequence because it defines an upper limit for the matrix molecular weight and thus a miscibility region in which the shell-only particles are miscible in linear polymer melts.

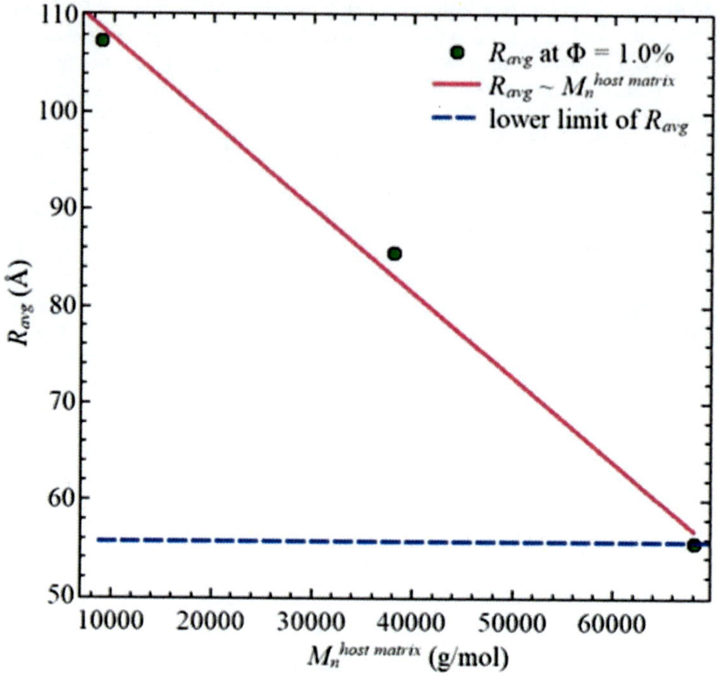

Fig. 4.10 Average radius, R_{avg}, vs. number molecular weight of the linear host matrix, $M_n^{host\ matrix}$, including the lower limit of the average radius (dashed blue line), based on the density of pure deuterated PDMS. *(Bichler, K. J., Jakobi, B., Huber, S. O., Gilbert, E. P., Schneider, G. J., Structural Analysis of Ultrasoft PDMS-g-PDMS Shell-Only Particles. Macromolecules 2020, 10.1021/acs.macromol.9b01598)*

The associated Flory–Huggins interaction parameter χ, obtained by fitting the data with Eq. (4.15), are very small but increase with increasing molecular weight of the linear host matrix. However, the values are always negative. A similar trend is reported for polybutadiene star polymers with labeled single arms mixed in two linear host matrices with increasing molecular weight [33]. In general, two components are only miscible if the χ-parameter is negative and additionally below the critical interaction parameter χ_{cr} [34, 35]. The critical parameter can be calculated as

$$\chi_{cr} = \frac{1}{2} \cdot \left(\frac{1}{\sqrt{DP_{Bottlebrush}}} + \frac{1}{\sqrt{DP_{linear\ chains}}} \right)^2 \tag{4.20}$$

with $DP_{Bottlebrush}$ and $DP_{Linear\ chains}$ being the degree of polymerization of the bottlebrushes and the linear chains of the host matrices, respectively. It follows the same trend as the experimentally determined values. Equation (4.20) assumes mixtures of linear polymers. Therefore, $DP_{Bottlebrush}$ can be approximated by twice the degree of polymerization of the side chains as the interacting segments, thus, $DP_{Bottlebrush} = 290$. The results are summarized in Table 4.7. All obtained values are

Table 4.7 χ-parameter as obtained from calculation and experiment for PDMS-g-PDMS bottlebrush polymer in linear host matrices with number average molecular weight, M_n.The standard deviations are smaller than 1% of the fit value *(Bichler, K. J., Jakobi, B., Huber, S. O., Gilbert, E. P., Schneider, G. J., Structural Analysis of Ultrasoft PDMS-g-PDMS Shell-Only Particles. Macromolecules 2020, 10.1021/acs.macromol.9b01598)*

χ-parameter	lin-h-PDMS9k	lin-h-PDMS38k	lin-h-PDMS68k
Experiment χ_{exp}	$-3.3 \cdot 10^{-3}$	$-1.1 \cdot 10^{-3}$	$-0.64 \cdot 10^{-3}$
Critical χ_{cr}	$1.01 \cdot 10^{-2}$	$4.94 \cdot 10^{-3}$	$4.17 \cdot 10^{-3}$

below χ_{cr} and are therefore consistent with our experimental observations of miscibility.

In the next step the concentration dependent evolution of the structure factor was explored. From its typical signatures, only the lowered forward scattering can be observed. In Fig. 4.8, a decrease of the intensity with increasing bottlebrush concentration is seen; remarkably, increasing the molecular weight lowers this decrease. At the highest molecular weight, the decrease of the intensity is less pronounced or even disappeared. Moreover, in all cases the typical correlation peak, associated with the structure factor is absent in our systems.

Both observations (intensity and peak) could have two possible reasons, (a) smearing effects inherent to the polymer itself or (b) reduction of the structure peak, known from star/linear polymer mixtures in solution.

Smearing is expected from the different contributions to the intensity in our samples. The main contrast is generated between the deuterated shell and the protonated linear chains of the host matrix. The contributions of the contrast between the protonated core and the protonated end groups of the deuterated side chains of our bottlebrush polymer are negligible for the form factor, $P(Q)$ (Fig. 4.7). However, they may influence the structure factor, $S(Q)$, because of the additional contrast between the perimeter of the shell and the core. Ultimately, this can cause a broadening of the correlation peak until it is completely smeared.

Second, a similar behavior is known from star/linear polymer mixtures in solution [25, 36–39]. If one assumes that the system represents the limiting case without solvent, the same underlying theory can be utilized and attempted to explain the observations based on the respective interaction potentials. In such a mixture, three different interaction potentials are present: star–star, V_{SS}, linear chain–linear chain, V_{CC}, and star–linear chain, V_{SC}.

The star–star interaction can be described with an ultrasoft potential, which can be written as [22, 30, 36, 37, 40, 41]:

$$\beta V_{SS}(r) = \left(\frac{5}{18}\right) f^{\frac{3}{2}} \begin{cases} \left(\frac{1}{1+\frac{\sqrt{f}}{2}} - \ln\left(\frac{r}{\sigma_S}\right)\right) & \text{for } r \leq \sigma_S \\ \frac{1}{1+\frac{\sqrt{f}}{2}}\left(\frac{\sigma_S}{r}\right)\exp\left(-\frac{\sqrt{f}(r-\sigma_S)}{2\sigma_S}\right) & \text{for } r > \sigma_S \end{cases} \tag{4.21}$$

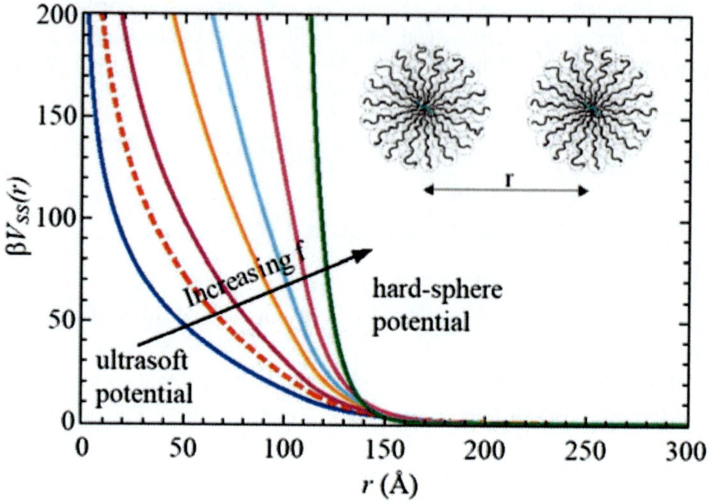

Fig. 4.11 Ultrasoft potential, $\beta V_{ss}(r)$, vs. distance, r, of spherical bottlebrush polymer with outer perimeter of $\sigma_s = 110$ Å for different number of side chains, f. From left to right: $f = 30$, 50, 75, 100, 150, and 500. The dotted red line represents our number of side chains $f = 40$. *(Bichler, K. J., Jakobi, B., Huber, S. O., Gilbert, E. P., Schneider, G. J., Structural Analysis of Ultrasoft PDMS-g-PDMS Shell-Only Particles. Macromolecules 2020, 10.1021/acs.macromol.9b01598)*

with f number of side chains, σ_S is the so-called corona diameter, equivalent to the star extension, r is the distance between two star centers, and $\beta = (k_B T)^{-1}$ where k_B is the Boltzmann constant and T is the temperature. This potential has a logarithmic dependence at short distances ($r \leq \sigma_S$) and decays exponentially at long distances ($r > \sigma_S$). With increasing f, the transition to the typical hard sphere potential is established (Fig. 4.11).

Additionally, the potential describing the interaction between two linear polymer chains, V_{CC}, is given by [36, 42]

$$\beta V_{CC}(r) = \left(\frac{5}{18}\right) f^{\frac{3}{2}} \begin{cases} \left(\frac{1}{2\tau^2\sigma_c^2} - \ln\left(\frac{r}{\sigma_c}\right)\right) & \text{for } r \leq \sigma_c \\ \frac{1}{2\tau^2\sigma_c^2} \exp\left(-\tau^2\left(r^2 - \sigma_c^2\right)\right) & \text{for } r > \sigma_c \end{cases} \tag{4.22}$$

This potential results from the star–star interaction potential, if the number of side chains is low ($f < 10$). The parameter σ_c measures the size of the polymer chain, r describes the distance between two linear chains by taking the middle monomeric unit as a center, and τ depends on the number of side chains and is a free parameter. From fitting of computer simulation data, $\tau = 1.03$ was determined for the case of a two-armed star polymer ($f = 2$), which reflects a linear chain [42].

The star–linear chain interaction can be written as [36]:

$$\beta V_{SC}(r) = \begin{cases} -\Theta(f)\ln\left(\frac{r}{\sigma_{sc}}\right) + K & \text{for } r \le \sigma_{sc} \\ v_0 \int \varphi(\mathbf{r}')_{\text{star}}\varphi(|\mathbf{r}-\mathbf{r}'|)_{\text{c}}\mathrm{d}\mathbf{r}' & \text{for } r > \sigma_{sc} \end{cases} \tag{4.23}$$

with $\sigma_{sc} = \frac{1}{2}(\sigma_s + \sigma_c)$, $\Theta(f) = \frac{5}{36}\left(\frac{1}{\sqrt{2}-1}\right)\left[(f+2)^{\frac{3}{2}} - \left(f^{\frac{3}{2}} + 2^{\frac{3}{2}}\right)\right]$, K is a constant which can be evaluated by requiring continuity for $V_{SC}(r)$ and its first derivative at $r = \sigma_{sc}$, v_0 is the excluded volume parameter and $\varphi(r)_{\text{star}}$ and $\varphi(r)_c$ are the radial dependent density profiles evaluated from the blob picture of the Daoud–Cotton model [6].

The effective interaction potential of a star polymer dispersed in linear chains, resulting from the three different interaction potentials in the system is derived by Camargo et al. [36, 43, 44] and Mayer et al. [38].

In such a system, once linear chains are added, the pure star–star correlation is reduced due to the linear chains, which are occupying space between overlapping stars. This results in a decrease in the internal fluctuations. As seen in scattering diagrams, this effect is connected with a drastic decrease of the structure peak, $S(Q)$, compared to a pure star polymer solution at higher concentration [25].

The reduction in intensity of the correlation peak arises from a depletion-like phenomenon with the linear chains as the depletion agent. Further details can be found in the recent work by Stiakakis et al. [45] and Mayer et al. [38]. The depletion effect was found to be extremely strong in melt compared to solvent for polymer nanocomposites and polymer/nanoparticle blends near a substrate [46–48]. The resulting osmotic force from the linear chains acts on the star polymers, especially seen by the changing size, depending on the molecular weight of the linear host matrix. For mixtures, with linear chains of approximately the same molecular weight, M_n, as one single arm, the star polymer swells due to interpenetration with the arms. If the molecular weight of the linear chains increases, the star polymer shrinks induced by the osmotic pressure. Additionally, it can be hypothesized that with increasing molecular weight, the osmotic pressure exerted on the star polymer increases as well; this is supported by the decreasing size of the bottlebrush polymers by increasing the molecular weight of the linear host matrix. With increasing concentration of the bottlebrush polymers in the highest molecular weight host matrix, the osmotic pressure forces the bottlebrushes closer together which subsequently leads to a formation of another structural level with non-spherical symmetry, which can be seen as the start of cluster formation [31, 49]. Similar behavior is known from polymer grafted nanoparticles. For polymer nanocomposites with $M_{\text{matrix}} \gg 2 \cdot M_{\text{graft}}$ the particle shell collapses and, subsequently, with increasing concentration of the polymer grafted nanoparticles, aggregation occurs [50, 51].

In solution a concentration dependence of the bottlebrush size was observed and predicted [29, 52]. This points the attention to the concentration dependent size. However, at high concentrations the analysis of the size is more difficult due to the dominant structure factor in the Guinier region. But the mid Q-region between

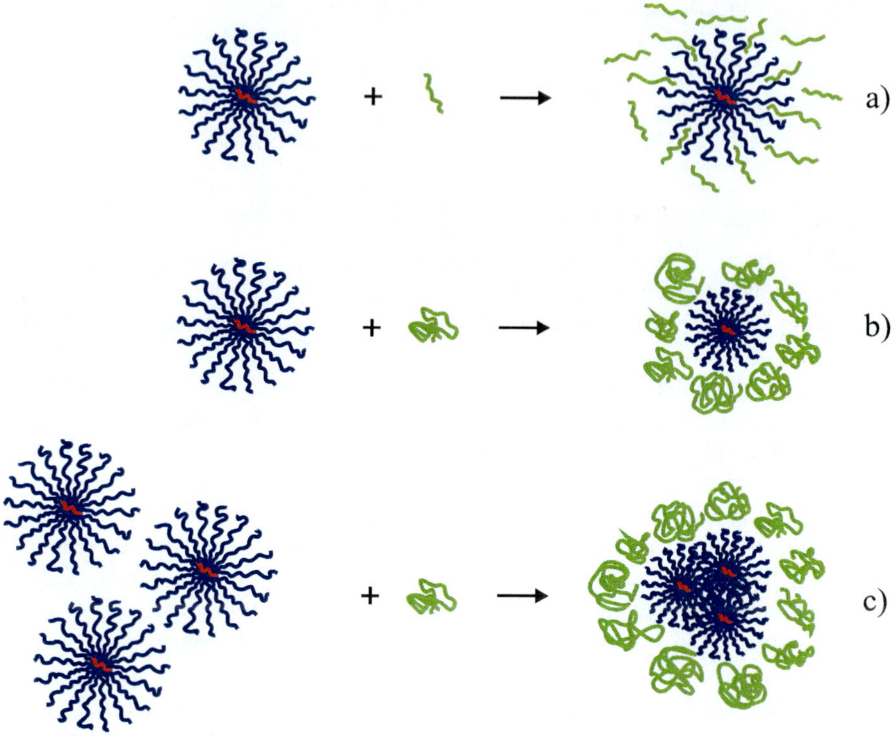

Fig. 4.12 Sketch of bottlebrush polymer mixed with linear chains of the same polymer: (**a**) bottlebrush polymer in matrix of low molecular weight linear chains, (**b**) bottlebrush polymer in matrix of high molecular weight linear chains, and (**c**) high concentration of bottlebrush polymers in matrix of high molecular weight linear chains. *(Bichler, K. J., Jakobi, B., Huber, S. O., Gilbert, E. P., Schneider, G. J., Structural Analysis of Ultrasoft PDMS-g-PDMS Shell-Only Particles. Macromolecules 2020, 10.1021/acs.macromol.9b01598)*

$Q = 0.05$ Å^{-1} and $Q = 0.08$ Å^{-1} can be analyzed regarding size changes. A change in radius should shift the measured scattering towards higher or lower Q-values for shrinking or expanding radius, respectively, as illustrated by the simulation of different radii (cf. Appendix A, Fig. A.2). Based on this assumption it can be concluded that the radius at this length scale is almost concentration independent, as even a change by 5% is directly observable.

As a short summary, all the findings are illustrated pictorially in Fig. 4.12.

4.2.3 Bottlebrush Polymers in Good Solvent: Shape Transition

The samples described hereafter are particularly interesting for studying the dynamical behavior under the influence of the shape transition from a more spherical into

Table 4.8 Weight average molecular weight, M_w, number average molecular weight, M_n, degree of polymerization, DP, polydispersity, M_w/M_n, and the grafting density of the three samples *(Jakobi, B., Bichler, K. J., Sokolova, A., Schneider, G. J., Dynamics of PDMS-g-PDMS Bottlebrush Polymers by Broadband Dielectric Spectroscopy. Macromolecules 2020, 10.1021/acs.macromol.0c01277)*

Name	M_w (g/mol)	M_n (g/mol)	DP	M_w/M_n	Grafting density (%)
Side chain short	298	298	4	1.0	89
Side chain middle	2300	1800	24	1.2	30
Side chain long	12,800	11,500	156	1.1	41
Backbone-1	17,500	16,500	275	1.1	–
Backbone-2	15,700	13,500	225	1.2	–

an elongated morphology. For the dynamical studies, dielectric spectroscopy and quasi-elastic neutron scattering have been performed. Additionally, results based on small-angle neutron scattering, are presented here.

The selected sample system consists of three samples, which differ in their side chain length, whereby the length of the backbone is, within the possibilities of the synthesis, kept constant. Based on the size ratio from backbone to side chain, the samples perform the shape transition from spherical/ellipsoid into elongated bottle-brush polymers. Small-angle neutron scattering is used to support this statement. The characteristic parameters of each sample, together with the grafting density, determined by GPC are summarized in Table 4.8.

Here the following matches have been done. The short side chains are grafted on backbone-1, while the other two side chains, middle and long, are attached to backbone-2.

A low concentration, $\Phi = 0.5\%$, of the bottlebrush polymers in deuterated cyclohexane is measured in order to identify the pure form factor, $P(Q)$. The presentation here follows chronological order, starting with the sample having the shortest side chains, $M_n^{\text{side chain}} = 298$ g/mol and finishes with the long side chain sample, $M_n^{\text{side chain}} = 11,500$ g/mol.

The first sample, PDMS-g-PDMS bottlebrush with the shortest side chains, $M_n^{\text{side chain}} = 298$ g/mol, displayed in Fig. 4.13, shows two kinks along with a power law of $Q^{-1.7}$. The power law in the transition region from Guinier (low Q-values) to the Porod region (intermediate Q-values) indicates an elongated shape. Additionally, this power law gives information about the solvent quality with the relationship to the Flory exponent, ν, of $Q^{-1/\nu}$. While a slope of Q^{-2} is characteristic for Θ- or melt condition, the apparent slope of $Q^{-1.7}$, with Flory exponent of $\nu = 0.588$, indicates good solvent condition [53]. Since PDMS is very flexible, the bottlebrush polymer is rather wormlike and semiflexible instead of stiff and rigid [54].

The two kinks in the scattering curve are connected to the overall dimension of the wormlike bottlebrush polymer. In the low Q-region, the kink is related to the contour length, L, of the polymer while the second kink at higher Q-values is connected with the cross-section radius, R. At very high Q-values, the blob region is detectable with the characteristic power law $Q^{-1.7}$ as described earlier. This adds a

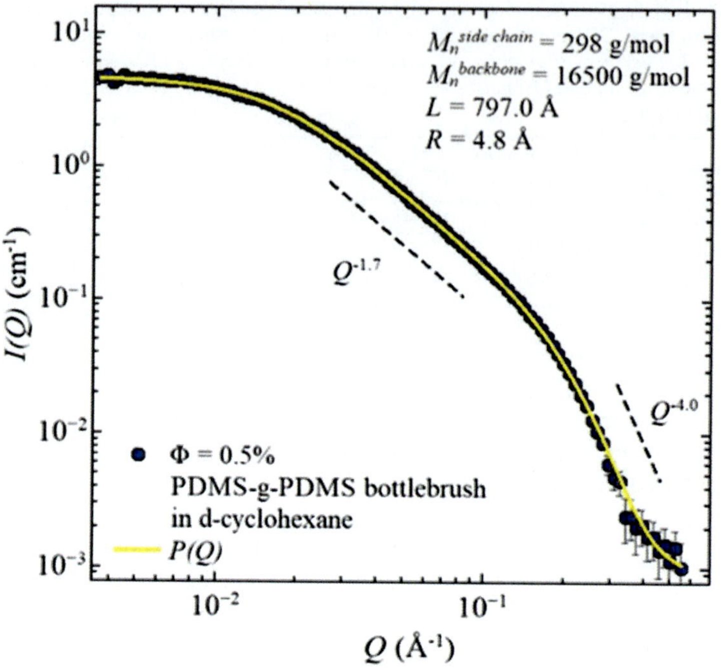

Fig. 4.13 Scattering intensity, $I(Q)$, vs. momentum transfer, Q, of a low concentration, $\Phi = 0.5\%$, of PDMS-g-PDMS bottlebrush polymer with $M_n^{\text{side chain}} = 298$ g/mol in d-cyclohexane. The solid lines are the data modeling with the form factor, $P(Q)$ (Eq. 4.8). *(Jakobi, B., Bichler, K. J., Sokolova, A., Schneider, G. J., Dynamics of PDMS-g-PDMS Bottlebrush Polymers by Broadband Dielectric Spectroscopy. Macromolecules 2020, 10.1021/acs.macromol.0c01277)*

further evidence of the good solvent condition. For this sample, the blob region is very small and therefore it is hard to extract a reliable blob size. Since the side chains only have four repeating units, a blob structure is unlikely and therefore the high Q-region may result from the background scattering.

By increasing the side chain length by a factor of around six to $M_n^{\text{side chain}} = 1800$ g/mol, while keeping the backbone length constant, the general shape is still rather elongated as seen in Fig. 4.14. The first kink in the low Q-region does not change, which is consistent with the same overall length of these two bottlebrush polymers.

However, the power law in the intermediate Q-region is less pronounced compared to the previous sample and the second kink moves to lower Q-values, which is compatible with an increased radius caused by increasing the molecular weight of the side chains. The less pronounced transition region could also be caused from polydispersity effects, which could be excluded due to the narrow distribution resulting from the synthesis of this sample. Furthermore, the blob region in the high Q increases as well, resulting in a well-developed power law of $Q^{-1.7}$ standing for the good solvent condition.

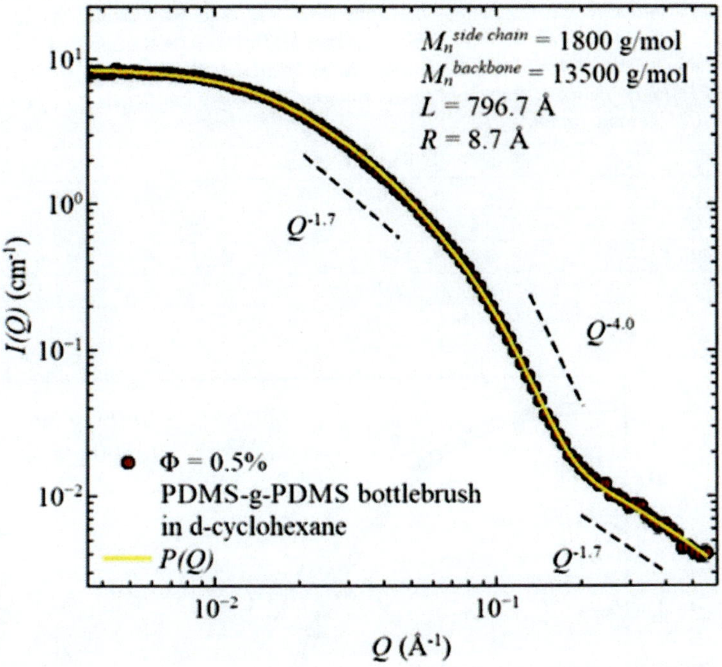

Fig. 4.14 Scattering intensity, $I(Q)$, vs. momentum transfer, Q, of a low concentration, $\Phi = 0.5\%$, of PDMS-g-PDMS bottlebrush polymer with $M_{\mathrm{n}}^{\text{side chain}} = 1800$ g/mol in d-cyclohexane. The solid lines are the data modeling with the form factor, $P(Q)$ (Eq. 4.8). *(Jakobi, B., Bichler, K. J., Sokolova, A., Schneider, G. J., Dynamics of PDMS-g-PDMS Bottlebrush Polymers by Broadband Dielectric Spectroscopy. Macromolecules 2020, 10.1021/acs.macromol.0c01277)*

The first two samples, $M_{\mathrm{n}}^{\text{side chain}} = 298$ g/mol and $M_{\mathrm{n}}^{\text{side chain}} = 1800$ g/mol can be well described by the form factor, $P(Q)$, for the elongated bottlebrush polymer as introduced and described in Sect. 4.1.2 and represented as the solid lines in Figs. 4.13 and 4.14. The results are summarized in Table 4.9.

As seen on the resulting fit parameters, the contour length, L, as well as the Kuhn length, ℓ_{k}, is, in this case, independent of the side chain length, only the radius and the blob size are affected form changing the molecular weight. Similar results have been obtained by Rathgeber et al. [13].

The third investigated sample has the longest side chain length of $M_{\mathrm{n}}^{\text{side chain}} = 11,500$ g/mol at the same backbone length. The associated scattering curve is shown in Fig. 4.15. Here, a direct transition from Guinier to Porod can be observed with no visible power law in between. This suggests a more spherical shape. Following a $Q^{-4.0}$ decay in the intermediate Q-region, the blob region starts at the high momentum transfers with the characteristic power law of $Q^{-1.7}$ implying a good solvent condition.

For describing the sample, the same model for spherical bottlebrush polymers, as introduced in Sect. 4.1.1 has been used and the best description is shown in Fig. 4.15.

Table 4.9 Contour length, L, radius, R, Kuhn length, ℓ_k, and blob size, ξ, of the elongated bottlebrush polymer with $M_n^{side\ chain} = 298$ g/mol and $M_n^{side\ chain} = 1800$ g/mol. All standard deviations are smaller than 1% *(Jakobi, B., Bichler, K. J., Sokolova, A., Schneider, G. J., Dynamics of PDMS-g-PDMS Bottlebrush Polymers by Broadband Dielectric Spectroscopy. Macromolecules 2020, 10.1021/acs.macromol.0c01277)*

	Short side chains	Middle side chains
Length L (Å)	797.0	796.7
Radius R (Å)	4.8	8.7
Kuhn length ℓ_k (Å)	43.6	43.6
Blob size ξ (Å)	2.7	7.5

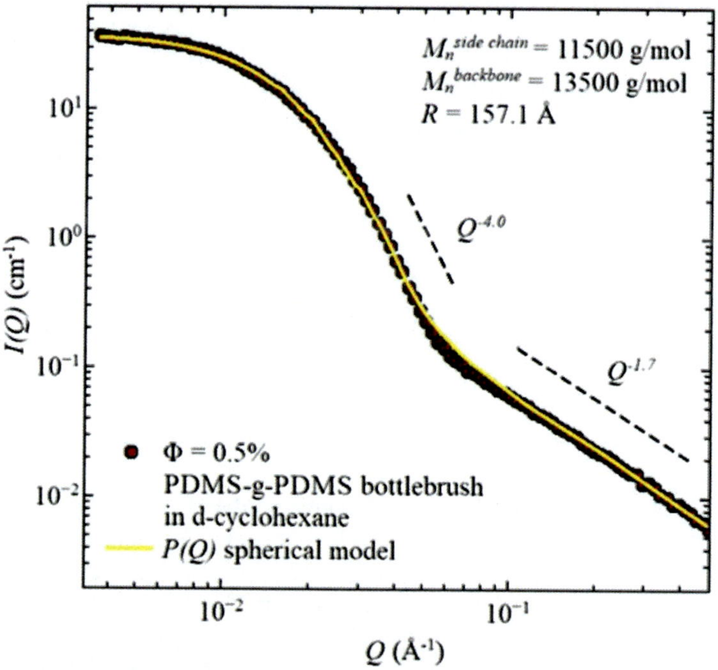

Fig. 4.15 Scattering intensity, $I(Q)$, vs. momentum transfer, Q, of the PDMS-g-PDMS bottlebrush polymer with $M_n^{side\ chain} = 11{,}500$ g/mol in d-cyclohexane. Full yellow line represents the form factor description with the model for spherical bottlebrush polymers

In the transition to the blob region, the model function does not describe the data properly. The low and the high Q-region separately are described well, with the resulting radius of $R = (157.1 \pm 1.5)$ Å with a polydispersity of around 19%. Since the backbone length is similar to the other two samples, one would assume that the radius of the sphere needs to be in the same size range as the obtained length, L, as obtained for the other two samples. Here, the radius is around ~1/5 of the length, i.e., $R \sim 1/5 \cdot L$. Therefore, the model for elongated bottlebrush polymers is also applied to this sample.

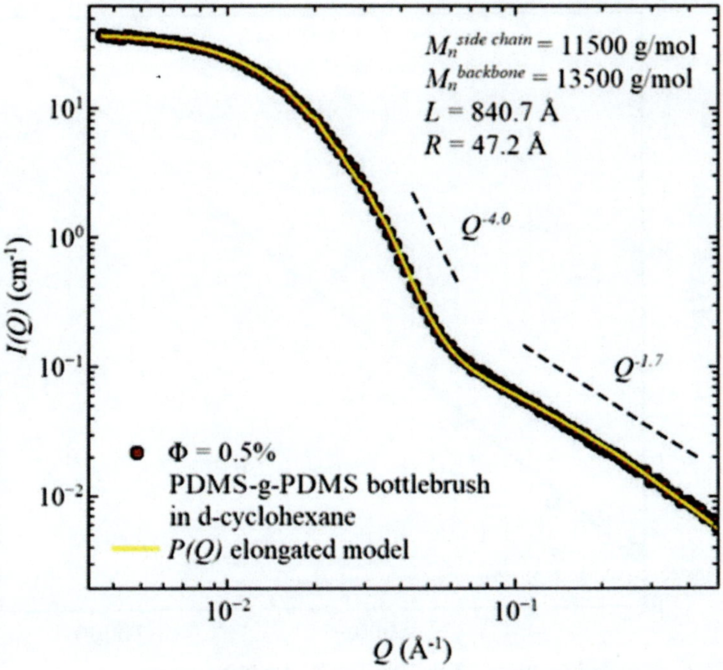

Fig. 4.16 Scattering intensity, $I(Q)$, vs. momentum transfer, Q, of the PDMS-g-PDMS bottlebrush polymer with the longest side chains in d-cyclohexane. Full yellow line represents the form factor description with the model for elongated bottlebrush polymers. *(Jakobi, B., Bichler, K. J., Sokolova, A., Schneider, G. J., Dynamics of PDMS-g-PDMS Bottlebrush Polymers by Broadband Dielectric Spectroscopy. Macromolecules 2020, 10.1021/acs.macromol.0c01277)*

Table 4.10 Contour length, L, radius, R, Kuhn length, ℓ_K, and blob size, ξ, of the PDMS-g-PDMS bottlebrush polymer with $M_n^{\text{side chain}} = 11,500$ g/mol. All standard deviations are smaller than 1% *(Jakobi, B., Bichler, K. J., Sokolova, A., Schneider, G. J., Dynamics of PDMS-g-PDMS Bottlebrush Polymers by Broadband Dielectric Spectroscopy. Macromolecules 2020, 10.1021/acs.macromol.0c01277)*

Name	Length L (Å)	Radius R (Å)	Kuhn length ℓ_K (Å)	Blob size ξ (Å)
Long	840.7	47.2	42.8	18.9

As seen in Fig. 4.16, this model describes the data better compared to the previous one. All obtained fitting values are summarized in Table 4.10.

By increasing the side chain length, the radius increases as expected, together with the blob size. Furthermore, in this case, the contour length, L, increases as well. The reason for that could be the little higher grafting density which forces the backbone to extend a bit more compared to the sample with middle side chains. Another explanation would be that the side chains can also graft at the chain ends of

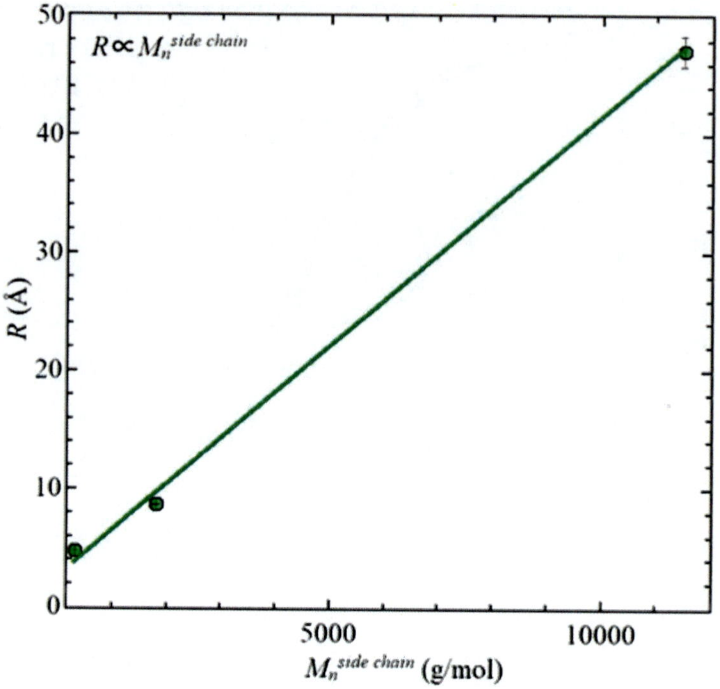

Fig. 4.17 Number average molecular weight of the side chains, $M_n^{\text{side chain}}$, vs. radius, R, of the cross-section from the elongated bottlebrush model. Solid green line represents a linear data description

the backbone which increases the molecular weight of the backbone by at least one time the molecular weight of the side chains and therefore the length increases. This could happen in all three samples. But an increase of the molecular weight of the backbone by the molecular weight from the short or the middle side chain length is not substantial.

Looking at the molecular weight dependence of the radius obtained from the elongated bottlebrush polymer model, a linear relationship of $R \propto M_n^{\text{side chain}}$, can be assumed (Fig. 4.17), at least for the molecular weights investigated here. This supports the assumption that the sample having the longest side chains can also be described in a good way by the model function for elongated bottlebrush polymers.

Since an elongated shape having a high polydispersity, in this case of 10%, cannot be distinguished from a polydisperse sphere, here 19%, by means of the scattering diagram [55], both model functions are suitable for the sample having the longest side chains of $M_n^{\text{side chain}} = 11,500$ g/mol. However, only the second model gives physically meaningful results.

References

1. J.S. Pedersen, Analysis of small-angle scattering data from colloids and polymer solutions: modeling and least-squares fitting. Adv. Colloid Interf. Sci. **70**, 171–210 (1997)
2. J. Pedersen, Small-angle scattering from surfactants and block copolymer micelles, in *Soft Matter Characterization*, (Springer, Berlin, 2008), pp. 191–233
3. R. Lund, V. Pipich, L. Willner, A. Radulescu, J. Colmenero, D. Richter, Structural and thermodynamic aspects of the cylinder-to-sphere transition in amphiphilic diblock copolymer micelles. Soft Matter **7**(4), 1491–1500 (2011)
4. T. Kawakatsu, *Statistical Physics of Polymers: An Introduction* (Springer, Berlin, 2013)
5. G. Beaucage, Small-angle scattering from polymeric mass fractals of arbitrary mass-fractal dimension. J. Appl. Crystallogr. **29**(2), 134–146 (1996)
6. M. Daoud, J.P. Cotton, Star shaped polymers: a model for the conformation and its concentration dependence. J. Phys. France **43**(3), 531–538 (1982)
7. W.D. Dozier, J.S. Huang, L.J. Fetters, Colloidal nature of star polymer dilute and semidilute solutions. Macromolecules **24**(10), 2810–2814 (1991)
8. G.S. Grest, L.J. Fetters, J.S. Huang, D. Richter, Star polymers: experiment, theory, and simulation. Adv. Chem. Phys. Polym. Syst. **94**, 67–163 (1996)
9. S. Gupta, M. Camargo, J. Stellbrink, J. Allgaier, A. Radulescu, P. Lindner, E. Zaccarelli, C.N. Likos, D. Richter, Dynamic phase diagram of soft nanocolloids. Nanoscale **7**(33), 13924–13934 (2015)
10. R. Lund, L. Willner, M. Monkenbusch, P. Panine, T. Narayanan, J. Colmenero, D. Richter, Structural observation and kinetic pathway in the formation of polymeric micelles. Phys. Rev. Lett. **102**(18), 188301 (2009)
11. M. Daoud, J.F. Joanny, Conformation of branched polymers. J. Phys. **42**(10), 1359–1371 (1981)
12. J.S. Pedersen, P. Schurtenberger, Cross-section structure of cylindrical and polymer-like micelles from small-angle scattering data. I. Test of analysis methods. J. Appl. Crystallogr. **29**(6), 646–661 (1996)
13. S. Rathgeber, T. Pakula, A. Wilk, K. Matyjaszewski, K.L. Beers, On the shape of bottle-brush macromolecules: systematic variation of architectural parameters. J. Chem. Phys. **122**(12), 124904 (2005)
14. W.-R. Chen, P.D. Butler, L.J. Magid, Incorporating intermicellar interactions in the fitting of SANS data from cationic wormlike micelles. Langmuir **22**(15), 6539–6548 (2006)
15. J.S. Pedersen, P. Schurtenberger, Scattering functions of semiflexible polymers with and without excluded volume effects. Macromolecules **29**(23), 7602–7612 (1996)
16. M. Rubinstein, R.H. Colby, *Polymer Physics* (OUP, Oxford, 2003)
17. H.G. Elias, *Makromoleküle: Chemische Struktur und Synthesen - Sechste, vollstandig überarbeitete Auflage* (Wiley, New York, 2009)
18. A.D. Jenkins, P. Kratochvil, R.G. Jones, E.S. Wilks, W.V. Metanomski, J. Kahovec, M. Hess, R. Stepto, T. Kitayama, T. Renner, *Compendium of Polymer Terminology and Nomenclature: IUPAC Recommendations 2008* (Royal Society of Chemistry, London, 2009)
19. L. Willner, O. Jucknischke, D. Richter, J. Roovers, L.-L. Zhou, P. Toporowski, L. Fetters, J. Huang, M. Lin, N. Hadjichristidis, Structural investigation of star polymers in solution by small-angle neutron scattering. Macromolecules **27**(14), 3821–3829 (1994)
20. G.S. Grest, K. Kremer, T.A. Witten, Structure of many arm star polymers: a molecular dynamics simulation. Macromolecules **20**(6), 1376–1383 (1987)
21. S.O. Huber, Microscopic view on structure and dynamics in polymer nanocomposites, Dissertation, Louisiana State University, 2019
22. C.N. Likos, H. Löwen, M. Watzlawek, B. Abbas, O. Jucknischke, J. Allgaier, D. Richter, Star polymers viewed as ultrasoft colloidal particles. Phys. Rev. Lett. **80**(20), 4450–4453 (1998)
23. F.J. Rogers, D.A. Young, New, thermodynamically consistent, integral equation for simple fluids. Phys. Rev. A **30**(2), 999 (1984)

24. L. Willner, O. Jucknischke, D. Richter, B. Farago, L.J. Fetters, J.S. Huang, Ordering phenomena of star polymers in solution by SANS. Europhys. Lett. **19**(4), 297 (1992)
25. B. Lonetti, M. Camargo, J. Stellbrink, C.N. Likos, E. Zaccarelli, L. Willner, P. Lindner, D. Richter, Ultrasoft colloid-polymer mixtures: structure and phase diagram. Phys. Rev. Lett. **106**(22), 228301 (2011)
26. B. Jakobi, Silica and siloxane model systems, Dissertation, Louisiana State University, 2019
27. K.J. Bichler, B. Jakobi, S.O. Huber, E.P. Gilbert, G.J. Schneider, Structural analysis of ultrasoft PDMS-g-PDMS shell-only particles. Macromolecules **53**(1), 78–89 (2020)
28. V. Arrighi, S. Gagliardi, A.C. Dagger, J.A. Semlyen, J.S. Higgins, M.J. Shenton, Conformation of cyclics and linear chain polymers in bulk by SANS. Macromolecules **37**(21), 8057–8065 (2004)
29. S. Bolisetty, S. Rosenfeldt, C.N. Rochette, L. Harnau, P. Lindner, Y. Xu, A.H. Muller, M. Ballauff, Interaction of cylindrical polymer brushes in dilute and semi-dilute solution. Colloid Polym. Sci. **287**(2), 129–138 (2009)
30. C.N. Likos, Effective interactions in soft condensed matter physics. Phys. Rep. **348**(4–5), 267–439 (2001)
31. A. Wilk, S. Huissmann, E. Stiakakis, J. Kohlbrecher, D. Vlassopoulos, C.N. Likos, G. Meier, J.K. Dhont, G. Petekidis, R. Vavrin, Osmotic shrinkage in star/linear polymer mixtures. Eur. Phys. J. E Soft Matter **32**(2), 127–134 (2010)
32. J. Kalb, D. Dukes, S.K. Kumar, R.S. Hoy, G.S. Grest, End grafted polymernanoparticles in a polymeric matrix: effect of coverage and curvature. Soft Matter **7**(4), 1418–1425 (2011)
33. R.W. Richards, Influence of architecture on arm dimensions and interaction parameters in polybutadiene star polymers. Macromolecules **32**(3), 880–891 (1999)
34. B. Andersen, Investigations on environmental stress cracking resistance of LDPE/EVA blends, Dissertation, Martin-Luther-Universität Halle-Wittenberg, Halle-Wittenberg, 2004
35. E. Manias, L.A. Utracki, Thermodynamics of polymer blends, in *Polymer Blends Handbook*, ed. by L. A. Utracki, C. A. Wilkie, (Springer Netherlands, Dordrecht, 2014), pp. 171–289
36. M. Camargo, C.N. Likos, Unusual features of depletion interactions in soft polymer-based colloids mixed with linear homopolymers. Phys. Rev. Lett. **104**(7), 078301 (2010)
37. C.N. Likos, Soft matter with soft particles. Soft Matter **2**(6), 478–498 (2006)
38. C. Mayer, C.N. Likos, A coarse-grained description of star−linear polymer mixtures. Macromolecules **40**(4), 1196–1206 (2007)
39. E. Stiakakis, G. Petekidis, D. Vlassopoulos, C.N. Likos, H. Iatrou, N. Hadjichristidis, J. Roovers, Depletion and cluster formation in soft colloid—polymer mixtures. Europhys. Lett. **72**(4), 664 (2005)
40. C.N. Likos, C. Mayer, E. Stiakakis, G. Petekidis, Clustering of soft colloids due to polymer additives. J. Phys. Condens. Matter **17**(45), S3363–S3369 (2005)
41. M. Watzlawek, H. Löwen, C.N. Likos, The anomalous structure factor of dense star polymer solutions. J. Phys. Condens. Matter **10**(37), 8189–8205 (1998)
42. A. Jusufi, J. Dzubiella, C.N. Likos, C. von Ferber, H. Löwen, Effective interactions between star polymers and colloidal particles. J. Phys. Condens. Matter **13**(28), 6177–6194 (2001)
43. M. Camargo, S.A. Egorov, C.N. Likos, Cluster formation in star-linear polymer mixtures: equilibrium and dynamical properties. Soft Matter **8**(15), 4177–4184 (2012)
44. M. Camargo, C.N. Likos, Phase separation in star-linear polymer mixtures. J. Chem. Phys. **130**(20), 204904 (2009)
45. E. Stiakakis, D. Vlassopoulos, C.N. Likos, J. Roovers, G. Meier, Polymer-mediated melting in ultrasoft colloidal gels. Phys. Rev. Lett. **89**(20), 208302 (2002)
46. V. Padmanabhan, A.L. Frischknecht, M.E. Mackay, Effect of chain stiffness on nanoparticle segregation in polymer/nanoparticle blends near a substrate. Macromol. Theory Simul. **21**(2), 98–105 (2012)
47. J.B. Hooper, K.S. Schweizer, T.G. Desai, R. Koshy, P. Keblinski, Structure, surface excess and effective interactions in polymer nanocomposite melts and concentrated solutions. J. Chem. Phys. **121**(14), 6986–6997 (2004)

48. V. Ganesan, A. Jayaraman, Theory and simulation studies of effective interactions, phase behavior and morphology in polymer nanocomposites. Soft Matter **10**(1), 13–38 (2014)
49. D. Vlassopoulos, E. Stiakakis, M. Kapnistos, Model soft colloids out of equilibrium: glass-like and re-entrant transitions. Rheol. Rev, 179–252 (2007)
50. M. Rizk, M. Krutyeva, N. Lühmann, J. Allgaier, A. Radulescu, W. Pyckhout-Hintzen, A. Wischnewski, D. Richter, A small-angle neutron scattering study of a soft model nanofiller in an athermal melt. Macromolecules **50**(12), 4733–4741 (2017)
51. X. Wang, V.J. Foltz, M. Rackaitis, G.G.A. Böhm, Dispersing hairy nanoparticles in polymer melts. Polymer **49**(26), 5683–5691 (2008)
52. J. Paturej, T. Kreer, Hierarchical excluded volume screening in solutions of bottlebrush polymers. Soft Matter **13**, 8534–8541 (2017)
53. R. Sigel, Form factor for distorted semi-flexible polymer chains. Soft Matter **14**(5), 742–753 (2018)
54. I.W. Hamley, *Block Copolymers in Solution: Fundamentals and Applications* (Wiley, New York, 2005)
55. R. Hosemann, S.N. Bagchi, *Direct Analysis of Diffraction by Matter* (North-Holland, Amsterdam, 1962)

Chapter 5
Dynamics

Abstract The dynamical behavior of PDMS-g-PDMS bottlebrush polymers has been investigated by several techniques to cover a broad time range, i.e., from pico to micro seconds. Reduced segmental relaxation times of bottlebrush polymers compared to their single linear side chains associated with the grafting process have been extracted from dielectric spectroscopy results. Entanglement free dynamics for bottlebrush polymers even up to high molecular weights have been shown by the mean square displacement of fast field cycling experiments. To relate the dynamics with length scales associated with different positions of the bottlebrush polymers, quasi-elastic neutron scattering (QENS) revealed at short times side chain length independent methyl group rotations, confined on a threefold jump. At the longest accessible times by QENS the segmental dynamics is tracked. Hereby, a strong dependence of the relaxation time on the side chain length and the momentum transfer was found, which suggest a non-Gaussian behavior of the dynamics in the short-time region. Based on the QENS results, a new approach towards a master curve of the mean square displacement was developed, considering the separate contributions from methyl groups and segments.

This chapter describes the dynamics of different bottlebrush polymer melts. At first the spherical PDMS-g-PDMS bottlebrush was investigated together with a comparison of different architectures. This has been done by dielectric spectroscopy and fast field cycling relaxometry. As a second part, the dynamical behavior of PDMS-g-PDMS bottlebrush polymers, performing the shape transition, is explored by dielectric spectroscopy and quasi-elastic neutron scattering. All systems have been measured in the pure melt, i.e., without any solvent.

K.J. Bichler, *Morphology and Dynamics of Bottlebrush Polymers*, Springer Theses,
https://doi.org/10.1007/978-3-030-83379-4_5

5.1 Data Analysis

5.1.1 Dielectric Spectroscopy

The PDMS based bottlebrush polymers have only a dipole moment perpendicular to the polymer chain, both in the backbone and the side chains (Fig. 2.6), and are thus type B polymers. Therefore, by using dielectric spectroscopy, only the segmental dynamics are probed.

For the description of relaxation processes, the Debye model is the simplest approach. This leads to the complex dielectric function

$$\epsilon^*(\omega) = \epsilon_\infty + \frac{\Delta\epsilon}{1 + i\omega\tau_D} \tag{5.1}$$

with τ_D as the characteristic relaxation time of the system [1]. Most of the time, the observed processes are non-ideal and the Debye model cannot be used. Commonly, the empirical Havriliak–Negami function (HN), an adaption of the Debye model for non-ideal, i.e., asymmetric, relaxation processes, is used to describe the dielectric data in the frequency domain [2].

$$\epsilon^*_{\mathrm{HN}}(\omega) = \epsilon_\infty + \frac{\Delta\epsilon}{\left(1 + (i\omega\tau_{\mathrm{HN}})^\beta\right)^\gamma} \tag{5.2}$$

Here, β and γ are the shape parameters, describing the asymmetric broadening of the relaxation peak with the restrictions of $0 < \beta \leq 1$ and $0 < \gamma\beta \leq 1$. This complex function can also be split in a real and an imaginary part

$$\begin{aligned} \epsilon'_{\mathrm{HN}}(\omega) &= \epsilon_\infty + \Delta\epsilon\, r(\omega) \cos\left[\gamma\psi(\omega)\right] \\ \epsilon''_{\mathrm{HN}}(\omega) &= \Delta\epsilon\, r(\omega) \sin\left[\gamma\psi(\omega)\right] \end{aligned} \tag{5.3}$$

with

$$\begin{aligned} r(\omega) &= \left[1 + 2(\omega\tau_{\mathrm{HN}})^\beta \cos\left(\frac{\beta\pi}{2}\right) + (\omega\tau_{\mathrm{HN}})^{2\beta}\right]^{-\frac{\gamma}{2}} \\ \psi(\omega) &= \arctan\left[\frac{\sin\left(\frac{\beta\pi}{2}\right)}{(\omega\tau_{\mathrm{HN}})^{-\beta} + \cos\left(\frac{\beta\pi}{2}\right)}\right] \end{aligned} \tag{5.4}$$

This results in the relaxation time τ_{HN} characteristic for using the Havriliak–Negami function [1]. However, τ_{HN} does not represent the "real" relaxation time τ_s, it is shifted a little bit due to the asymmetry of the relaxation peak. In order to correct this, the following equation needs to be applied [3–6].

$$\frac{1}{\tau_s} = \frac{1}{\tau_{HN}} \left[\sin\left(\frac{\beta\pi}{2+2\gamma} \right) \right]^{\frac{1}{\beta}} \left[\sin\left(\frac{\beta\gamma\pi}{2+2\gamma} \right) \right]^{-\frac{1}{\beta}} \tag{5.5}$$

The temperature dependence of the segmental relaxation can be described with the VFT equation.

5.1.2 *Fast Field Cycling (FFC) Relaxometry*

Each temperature gives 3.5 decades of data points. In order to extend this narrow frequency window, the Time–Temperature Superpositions (TTS) principle or the equivalent Frequency–Temperature Superposition (FTS) principle is applied in frequency direction (i.e., only shifted in *x*-direction) to obtain the susceptibility master curve representation, $\chi''(\omega)$ vs. ω, as explained in Sect. 2.4.

$$\chi''_{DD} = K[\chi''(\omega) + 2\chi''(2\omega)] \tag{5.6}$$

The peak developing in the master curve can be described, by using the imaginary part of the HN function, $\chi''(\omega) = \mathrm{HN}(\omega)$, and taking $\beta = 1$ to avoid any contribution of polymer dynamics.

$$\chi''_{DD} = K[\mathrm{HN}(\omega) + 2\mathrm{HN}(2\omega)] \tag{5.7}$$

The resulting time constant based on the HN function needs to be corrected equivalent to dielectric spectroscopy. Subsequently, the final relaxation time constant, τ_s, is associated with the reference temperature of the master curve. The shift factors $a_T(T)$ used to create the master curve can be transformed to the relaxation time constants, τ, for every measured temperature by multiplication with the reference value τ_s. Plotting the master curve in the reduced frequency representation, $\chi''_{DD}(\omega)$ vs. $\omega\tau_s$, allows to compare different molecular weights [7]. The absolute mean square displacement is received by taking the cosine transformation of the master curve in the relaxation rate representation and applying Eq. (2.53) [8]. Furthermore, plotting $R_1(\omega)$ vs. $\omega^{0.5}$ for every temperature, Eq. (2.56) can be used to extract the temperature dependent diffusion coefficient and combined with Eq. (2.57) the radius of gyration.

In order to compare the relaxation times with those, obtained from dielectric spectroscopy, both data sets can be described simultaneously with the VFT equation.

5.1.3 Quasi-Elastic Neutron Scattering (QENS)

Depending on the measurement temperature of our samples, different processes are within the available frequency/time window. Usually, the so-called Elastic Fixed Window (EFW) scan is done prior to the quasi-elastic scans (Fig. 5.1), to select the temperatures of interest. For this experiment, the Doppler drive at the backscattering instrument is stopped and thus the intensity at $\Delta E = 0$ is obtained. In order to increase the signal-to-noise ratio to a good value, often the Q-average intensity is plotted as a function of temperature. This representation enables a direct visibility of plateau values, decreasing intensities and transitions between different regions. The initial plateau at low temperatures indicates no visible dynamics in the experimental time window of the instrument, while the final value, ~0, points to dynamics being too fast for the instrumental time range, or not within the Q-range.

Here it is to mention that the EFW scan was only used for a rough estimation of the measurement temperatures, but not for further data analysis. Due to the fast cooling process, which is needed to avoid crystallization of the samples, the temperatures represented in the EFW scan are the set-values, which are not identical with the actual temperature of the sample at the respective point. The heat conductivity of the sample can cause a delay, in the temperature equilibration, resulting in inaccurate temperature information. Before the quasi-elastic scans, the temperature of the sample was kept constant for a sufficient time, until temperature equilibrium has been reached to ensure a stable temperature with high accuracy.

The measured intensity, $S_{\text{exp}}(Q,\omega)$, is a convolution of the pure sample scattering, $S(Q,\omega)$, with the instrument specific resolution, $R(Q,\omega)$, i.e., $S_{\text{exp}}(Q,\omega) = S(Q,\omega) \otimes R(Q,\omega)$. This makes it difficult to combine different time windows from different instruments. However, for polymers a broad time/frequency window is needed to follow the dynamical processes, which is only possible to obtain by combining several instruments. For that, Fourier transformation of the experimental

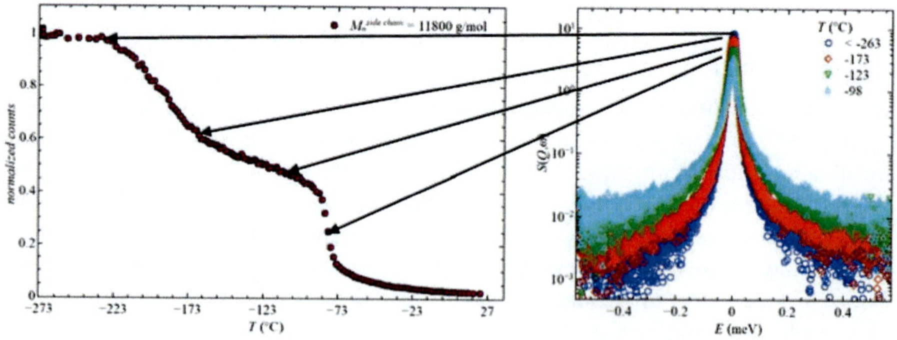

Fig. 5.1 Elastic fixed window scan (EFW) of one sample, showing exemplary the two different processes available in our sample systems as determined by a step in the EFW scan. The temperature represents the set temperature, the actual temperature depends on the heat conductivity of the material, therefore, differences between set temperature and actual temperature occur

data into the time domain turns the convolution into a multiplication and enables to combine the measurement range of the three instruments.

$$S_{\mathrm{exp}}(Q,t) = S(Q,t) \cdot R(Q,t) \tag{5.8}$$

This simplifies the way to obtain the intermediate scattering function, $S(Q,t)$, which is then a division of the Fourier transformed measurement signal, $S_{\mathrm{exp}}(Q,\omega)$, by the Fourier transformed instrumental resolution, $R(Q,\omega)$ [9, 10].

$$S(Q,t) = \frac{\int S_{\mathrm{exp}}(Q,\omega)\,\exp(i\omega t)\,\mathrm{d}\omega}{\int R(Q,\omega)\,\exp(i\omega t)\,\mathrm{d}\omega} \tag{5.9}$$

The resulting $S(Q,t)$ can be described with an appropriate model function.

For the combination, the IRIS data as well as the EMU data were shifted with the PELICAN data as reference. The need for the minimal shifting of the data can be explained by the structural heterogeneity inherent of the analyzed samples.

Based on the two distinct steps in the EFW scan, two processes are present in the available time window. The first decrease is associated with the methyl group reorientation. This process is the dominating part until $T = -23$ °C. The next step arises from the segmental relaxation which determines the measurement signal at high temperatures.

Methyl group reorientation, also known as localized motion, is a process geometrical restricted by the available jumping distance between different hydrogen atoms. This constraint can be well described by the so-called Elastic Incoherent Structure Factor (EISF). For describing the localized motions (represented as index ℓ), the used model function can be written as

$$S_\ell(Q,t) = A \cdot \exp\left(-\frac{1}{3}u^2Q^2\right) \cdot \left(\mathrm{EISF} + (1-\mathrm{EISF}) \cdot \exp\left(-\left(\frac{t}{\tau_\ell}\right)^{\beta_\ell}\right)\right) \tag{5.10}$$

with the Debye–Waller factor and a prefactor A. Since in polymers a huge number on different relaxation times exist, the Kohlrausch–Williams–Watts (KWW) function, $\exp\left(-\left(\frac{t}{\tau_l}\right)^{\beta_l}\right)$, is very convenient to use [9, 11–13]. In this context, τ_ℓ describes the relaxation time of the methyl groups and β_ℓ is the stretching exponent, resembling the width of the relaxation time distribution.

Typical for the methyl group reorientation is the so-called threefold jump introduced by Eq. (2.28) [13–15]. Here, the hydrogen atoms can jump between three equidistant positions on a circle with radius R (Fig. 5.2a). The threefold jump, represented by the EISF, is an oscillation around $\mathrm{EISF} = \frac{1}{3}$ for large Q-values

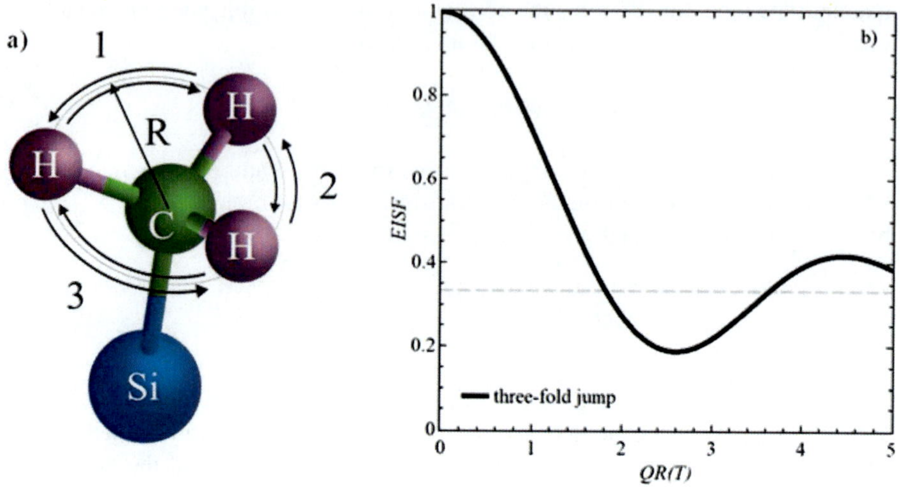

Fig. 5.2 Pictorial representation of the methyl group reorientation. (**a**) Schematic representation of a methyl group exemplary bonded to a silicon atom, showing the three options for the jumping process, including the radius R. (**b**) Elastic incoherent structure factor, EISF, vs. QR of the methyl group reorientation, i.e., threefold jump. (*Bichler, K. J., Jakobi, B., García Sakai, V., Klapproth, A., Mole, R. A., Schneider, G. J., Short Time Dynamics of PDMS-g-PDMS Bottlebrush Polymer Melts Investigated by Quasi-Elastic Neutron Scattering. Macromolecules 2020, 10.1021/acs.macromol.0c01846*)

(Fig. 5.2b). This assumption is also valid for our sample systems, as seen later in the data discussion.

At higher temperatures, $T \geq -23\ ^\circ\mathrm{C}$, the segmental relaxation is the measured quantity, whereby the Debye–Waller factor contributes to the measured signal as well. This relaxation is not restricted to any geometrical constraints, therefore, an additional EISF is not needed for the data description. The used model function is

$$S_s(Q,t) = A \cdot \exp\left(-\frac{1}{3}\langle u^2\rangle Q^2\right) \cdot \exp\left(-\left(\frac{t}{\tau_s}\right)^{\beta_s}\right) \tag{5.11}$$

with τ_s the segmental relaxation time and β_s the associated stretching parameter of the KWW function.

Both model functions have a large number of adjustable parameters. Therefore, to improve the stability of the fitting function reasonable assumptions are needed. For the low temperature function (Eq. 5.10) the atomistic mean square displacement $\langle u^2\rangle$ as well as the relaxation time of the localized motion τ_ℓ and the associated stretching parameter β_ℓ are independent of Q. In case of the Debye–Waller factor, the Q-dependence is already included as the Q^2 factor within the exponential function and therefore, $\langle u^2\rangle$ only depends on temperature. The relaxation time of the localized motion occurs naturally on a well restricted geometry and therefore, no significant Q-dependence should be detectable [15]. Furthermore, QENS experiments on

Table 5.1 Summary of all the Q-independent and dependent fitting parameters associated with the two model functions, low temperature, Eq. (5.10) and high temperature, Eq. (5.11)

Fitting function	Parameter	Q-independent	Q-dependent
Low temperature	A		x
	$\langle u^2 \rangle$	x	
	EISF		x
	τ_ℓ	x	
	β_ℓ	x	
High temperature	A		x
	$\langle u^2 \rangle$	x	
	τ_s		x
	β_s	x	

similar sample systems have not shown any substantial dependence on Q within the experimental error for the stretching parameter, β_ℓ [13, 16]. The elastic incoherent structure factor shows an explicit Q-dependence, which is determined separately.

In case of the segmental relaxation, described with the model function for high temperatures (Eq. 5.11), the relaxation time shows a pronounced Q-dependence [9, 17, 18]. Therefore, τ_s needs to be determined for each Q-value independently. Only the stretching parameter β_s and $\langle u^2 \rangle$ are treated to be independent of the momentum transfer. All the Q-independent and dependent parameters of the associated fitting function are summarized in Table 5.1.

All Q-values of one temperature are fitted with a so-called Global Fit. By means of that, all Q-values are simultaneously described with the respective model function, whereby the Q-independent parameters are shared over all Q-values. Shared means that all values do not depend on Q and should therefore be the same for all momentum transfers.

5.2 Results and Discussion

This section introduces the dynamics of different polymer architectures, i.e., bottlebrush and linear polymers, and investigates the changes in relaxation behavior arising from the variation of the shape.

5.2.1 Comparison of Polymers with Different Architectures

Within this subsection, the dynamics of a bottlebrush polymer melt, having a molecular weight of M_n~380 kg/mol, is compared with linear polymers of different lengths. First, the spherical PDMS-g-PDMS bottlebrush polymer (cf. Sect. 4.2.1) is compared with their respective single linear PDMS side chain. Second, a comparison

Table 5.2 Weight average molecular weight, M_w, number average molecular weight, M_n, and polydispersity, M_w/M_n, of the used linear PDMS and for comparison of the PDMS-g-PDMS bottlebrush (*Bichler, K. J., Jakobi, B., Schneider, G. J., Dynamical Comparison of Different Polymer Architectures – Bottlebrush vs. Linear Polymer. Macromolecules 2021, 10.1021/acs.macromol.0c02104*)

Name	M_w (g/mol)	M_n (g/mol)	M_w/M_n
Linear PDMS	380,000	370,000	1.03
PDMS-g-PDMS	431,000	385,000	1.12

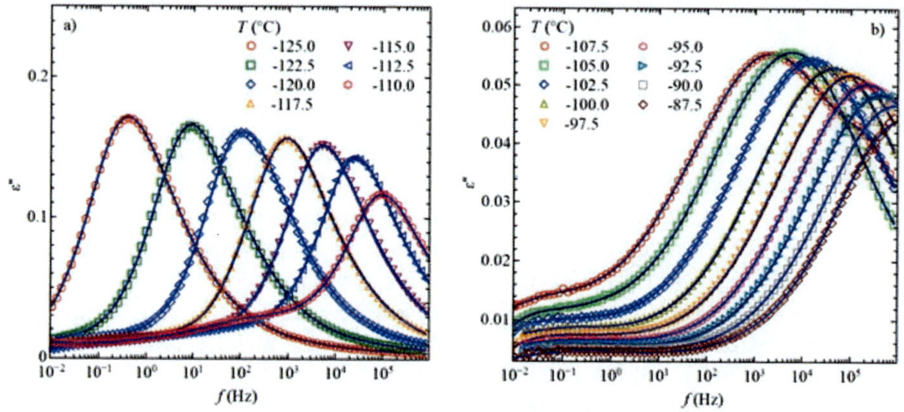

Fig. 5.3 Imaginary part of the permittivity, ϵ'', vs. frequency, f, for the spherical PDMS-g-PDMS bottlebrush polymer. (**a**) Segmental relaxation, α-relaxation and (**b**) the segmental relaxation influenced by the cold crystallization, α_c-relaxation. The solid lines are the best description using the Havriliak–Negami function *(Bichler, K. J., Jakobi, B., Schneider, G. J., Dynamical Comparison of Different Polymer Architectures – Bottlebrush vs. Linear Polymer. Macromolecules 2021, 10.1021/acs.macromol.0c02104)*

between bottlebrush and linear architecture is done, whereby both polymers have around the same M_n. Therefore, the molecular weight of the single linear PDMS side chains from the PDMS-g-PDMS bottlebrush polymer, as well as, for the whole bottlebrush can be found in Table 4.1 on page 64. For the second comparison, the molecular weight of the linear PDMS, can be found in Table 5.2. For comparison, the PDMS-g-PDMS bottlebrush (similar to Table 4.1) is also included.

5.2.1.1 Dielectric Spectroscopy of PDMS-g-PDMS Bottlebrush Polymer

Dielectric spectroscopy at different temperatures on the spherical PDMS-g-PDMS bottlebrush polymer is illustrated in Fig. 5.3 as dielectric permittivity, ϵ'', vs. frequency, f. Here two relaxation processes are visible. The first one starts at $T = -125$ °C up to $T = -110$ °C (Fig. 5.3a). In this region, the relaxation peak shifts to higher frequencies with increasing temperature, which is associated with a decrease in the relaxation strength, seen as a reduced height of the relaxation peak. After the first relaxation, the second one starts at $T = -107.5$ °C and lasts until

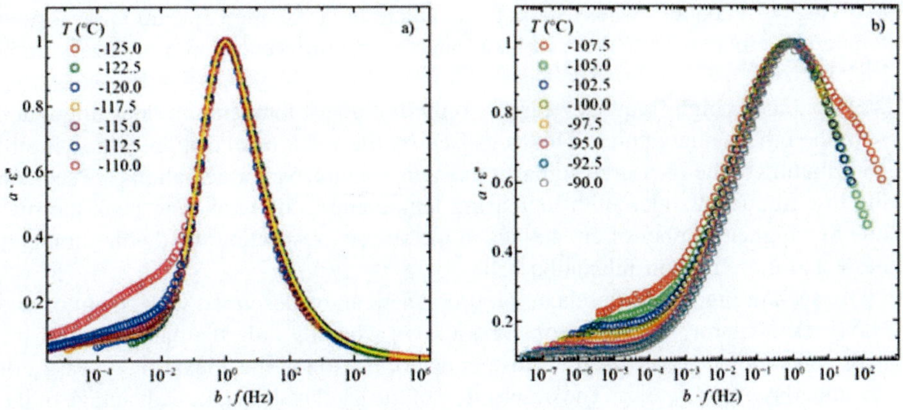

Fig. 5.4 Normalized imaginary part of the permittivity, $a \cdot \epsilon''$, vs. normalized frequency, $b \cdot f$, for the spherical PDMS-g-PDMS bottlebrush polymer. (**a**) Segmental relaxation, α-relaxation and (**b**) the segmental relaxation influenced by the cold crystallization, α_c-relaxation. The solid yellow line is the best description using the Havriliak–Negami function

$T = -87.5$ °C (Fig. 5.3b). Here the same behavior is seen. With increasing temperature, the peak position shifts to higher frequencies, connected with a decrease in the peak height. One main difference is seen at the first glance. While the first relaxation expands over almost six orders of magnitudes by increasing the temperature by $\Delta T = 15$ °C, the second relaxation only spreads over three decades by increasing the temperature by $\Delta T = 20$ °C. Therefore, the first relaxation has a stronger temperature dependence.

Based on literature, these two processes are typical for PDMS with linear architecture [19, 20]. The first relaxation is attributed with the segmental dynamics, known as α-relaxation (Fig. 5.3a), while the second process is a modified segmental relaxation which is influenced by the cold crystallization, the α_c-relaxation (Fig. 5.3b). In the available frequency window, dielectric spectroscopy measures the fluctuations of the permanent dipole moments, which are perpendicular to the polymer chain. This results in the segmental relaxation (Fig. 5.3a). Heating up the sample above the glass transition temperature, T_g, but still below the melting point, leads to the so-called cold crystallization, in case of PDMS. Since PDMS is a semi-crystalline polymer, in that temperature region, a coexistence of crystalline and amorphous regions with mobile chains is in the system. Therefore, the second process is named α_c-relaxation, indicating, that still the α-relaxation is measured, but now under the influence of the cold crystallization, i.e., a relaxation within a constraint amorphous phase, caused by the formation of crystallites (Fig. 5.3b) [19–21].

To compare the shape of the relaxation peaks of the different temperatures, each data set can be normalized in frequency and in permittivity, ϵ'', as seen in Fig. 5.4. The shape of the α-relaxation is almost independent of temperature (Fig. 5.4a). Only at the low frequency side, there are visible deviations coming from the onset of an evolving α_c-relaxation which starts to dominate the relaxation spectra upon further temperature increase. For the α_c-relaxation, stronger differences are visible as seen in Fig. 5.4b.

Here, at the high frequency region, only the lowest temperature does not match with the other temperatures. This is caused by the weak α-relaxation, which is still contributing to the measured signal at this temperature. Major deviations are seen on the low frequency side. With increasing temperature, the relaxation peak narrows due to a higher number of crystallites in the system, restricting the dipole moments more and more in their relaxation behavior.

As seen in Fig. 5.3 both relaxation processes can be described with a sum of two Havriliak–Negami (HN) functions (Eq. (5.3)), whereby only the parameters of the main relaxation peak are used for further determination of the relaxation times, τ_S, of the underlaying dynamics. The description of the α-relaxation takes advantage of the temperature independent shape, i.e., only the normalized data curve is described, as seen in Fig. 5.4a. Based on the normalization factors, especially the shift parameter in x-direction, the relaxation times, τ_S, can be determined. The resulting parameters for the best description of the dielectric spectroscopy data sets can be found in the Appendix.

While the spectra in Fig. 5.3 contain all the information, needed for the discussion, often it is more convenient to concentrate on the relaxation times, τ_S vs. $1000/T$ as seen in Fig. 5.5.

It contains the same information, i.e., the processes are becoming faster with increasing temperature. Furthermore, the cold crystallization is immediately visible, once the α-relaxation changes to the α_c-relaxation. Additionally, this representation allows a much easier description via the VFT equation. The resulting fit parameters are summarized in Table 5.3.

The PDMS-g-PDMS bottlebrushes are synthesized by the grafting-to-technique. This permits to study the dynamics of the single side chains, i.e., linear PDMS, separately, before attaching them to the backbone. In this way, the influences on the α-relaxation caused by the grafting process can be compared. Since the dielectric spectroscopy spectra, themself do not provide new information, they are displayed in the Appendix. The comparison of the relaxation times, τ_S, vs. $1000/T$, of the linear single PDMS side chains and the PDMS-g-PDMS bottlebrush polymer are illustrated in Fig. 5.6.

Clear differences in the relaxation times, by a factor of $5 - 10$, for the segmental relaxation are visible, whereas the α_c-relaxation shows only slight changes. In case of the bottlebrush polymer, both parts, i.e., backbone and side chains contribute to the measurement signal. Since the amount of the backbone is below 1% of the overall molecular weight (cf. Table 4.1), the majority of the dipole moments are located in the side chains and it can be assumed that the dielectric spectroscopy spectra are dominated by them. The main differences between bottlebrush polymer and single side chains are that in the former case the side chains are fixed with one end to the backbone and neighbored side chains create a confined environment, which results in strong intermolecular interactions between adjacent chains. These restrictions cause the slowdown of the α-relaxation for the bottlebrush polymer

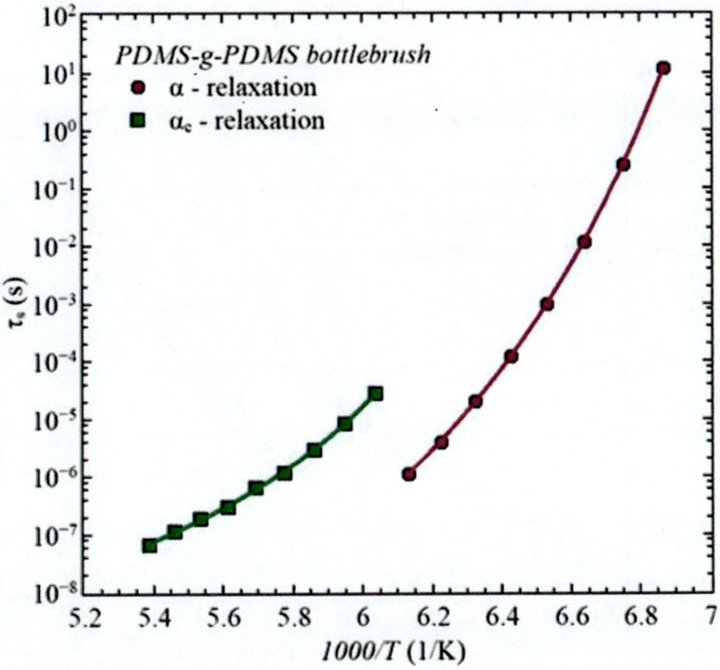

Fig. 5.5 Segmental relaxation times, τ_s, vs. 1000/T of the spherical PDMS-g-PDMS bottlebrush polymers. Solid lines are the best description with the VFT equation

Table 5.3 Fit parameter of the VFT equation for PDMS-g-PDMS bottlebrush, describing the temperature dependence of the relaxation time of the α- and α_c-relaxations including the limiting relaxation time at high temperatures, τ_∞, the fit parameter, A, and the ideal glass transition temperature, T_0

	τ_∞ (s)	A (K)	T_0 (K)
α-relaxation	$(5 \pm 1) \cdot 10^{-15}$	734 ± 9	125.9 ± 0.2
α_c-relaxation	$(3 \pm 2) \cdot 10^{-11}$	351 ± 47	140 ± 2

compared to the single side chains. Similar results have been obtained by Grigoriadis et al. [22] for PBiBEM-g-PBA bottlebrush polymers.

The relaxation time of the α_c-relaxation is approximately the same between the two structures. However, in case of the bottlebrush polymer this relaxation starts at higher temperatures compared to the single side chains. This can be attributed with a higher onset temperature of the cold crystallization of PDMS-g-PDMS bottlebrush polymer compared to linear PDMS. Linear polymers crystallize easier, compared to branched polymers, due to their regularity [23–25]. In case of bottlebrush polymers, the bulky side chains generate a kind of irregularity, or heterogeneity into the systems, which hinders the formation of crystallites. This could be the reason for the shifted α_c-relaxation in our sample systems.

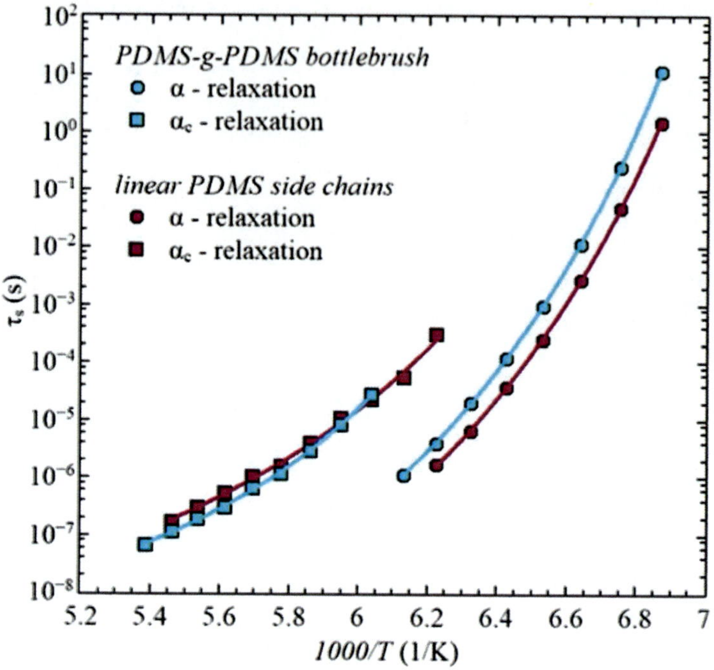

Fig. 5.6 Segmental relaxation times, τ_s, vs. 1000/T of the spherical PDMS-g-PDMS bottlebrush polymers compared to the single linear PDMS side chains. Solid lines are the best description with the VFT equation

Table 5.4 Glass transition temperatures, T_g, of PDMS-g-PDMS bottlebrush and linear PDMS side chains, determined by dielectric spectroscopy

Sample	T_g (°C)	T_g (K)
PDMS-g-PDMS bottlebrush	-127.7 ± 0.1	145.5 ± 0.1
Linear PDMS side chains	-129.9 ± 0.9	143.3 ± 0.9

Using the fit parameters for describing the VFT behavior of the segmental relaxation allows to determine the glass transition temperature, T_g, of both samples, i.e., bottlebrush and single side chains, as the temperature at which the relaxation time $\tau_s = 100$ s. This is an estimation established in literature and shows, in general, a good agreement with Differential Scanning Calorimetry (DSC) measurements [5, 26, 27]. The determined glass transition temperatures are summarized in Table 5.4 for both samples. Comparable literature values, determined by DSC, for linear PDMS are $T_g = -127.5$ °C and $T_g = -149.85$ °C to $T_g = -123.25$ °C [2, 28, 29].

Based on rheological experiments of heteropolymer bottlebrushes, T_g is mostly determined by the side chains, independent of the grafting density [30–32]. This statement can also be assumed in case of the PDMS-g-PDMS bottlebrush polymers.

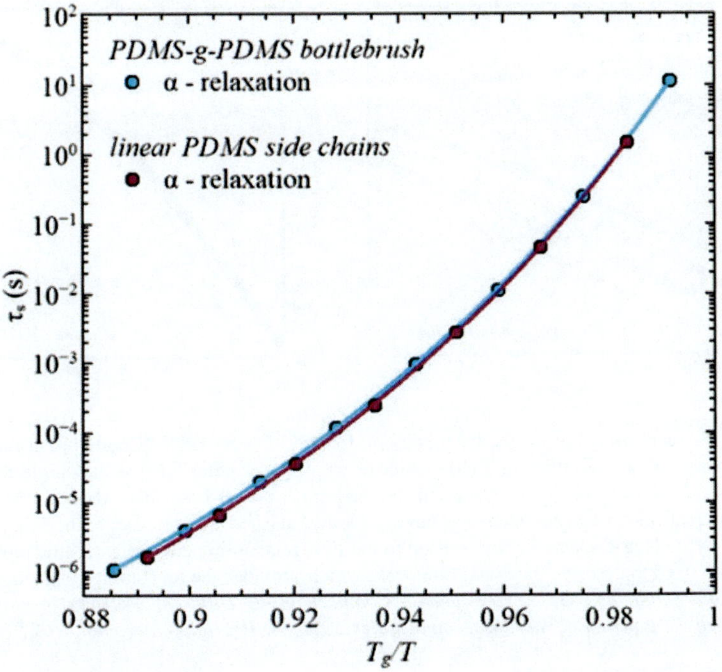

Fig. 5.7 Segmental relaxation times, τ_s, vs. T_g/T of the PDMS-g-PDMS bottlebrush polymer and the single linear PDMS side chains. The full lines are the best descriptions with the VFT equation

As seen in Table 5.4, there is a deviation of about $\Delta T = 2$ °C between the glass transition temperatures of the bottlebrush polymer and the linear PDMS. To analyze the segmental relaxation times without any effects of the different glass transition temperatures, it is common in literature to normalize the data against their respective T_g [33]. This is shown in Fig. 5.7 as segmental relaxation time, τ_s, vs. T_g/T. In the entire temperature range, both samples coincide and thus relate to an identical temperature behavior of the α-relaxation, relative to T_g. This is common for a large number of polymers. Polymers of different molecular weight, scaled to T_g show the same relaxation behavior and thus coincide.

5.2.1.2 Fast Field Cycling (FFC) Relaxometry of Bottlebrush and Linear Polymer

Fast field cycling (FFC) relaxometry experiments were performed on the two samples, described in Table 5.2, and previously used for dielectric spectroscopy.

The relaxation rates, $R_1(\omega)$, as a function of the frequency, ω, were measured at different temperatures. As shown in Fig. 5.8a, it is very common to transform the relaxation rates into a susceptibility representation, $\chi''(\omega) = R_1(\omega) \cdot \omega$, and plot it as a function of ω for several temperatures. The used frequency range is restricted, which

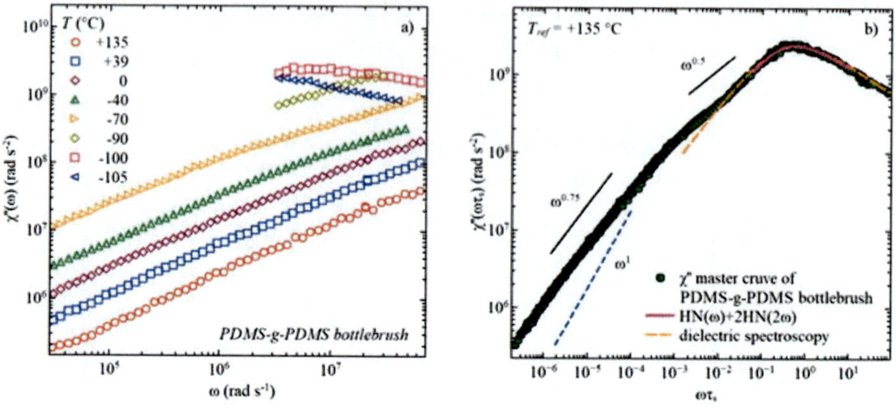

Fig. 5.8 Susceptibility, $\chi''(\omega)$ vs. frequency, ω for the PDMS-g-PDMS bottlebrush polymer. (**a**) Single measured temperatures and (**b**) master curve obtained after FTS of the single temperature measurements. Black solid lines represent the theoretical power laws. Blue dashed line represents the theoretical power law for Debye behavior. Orange dashed line represents the relaxation peak obtained by dielectric spectroscopy, shifted to the FFC relaxation peak and pink line represents the fit function used to describe the relaxation peak, which provides the relaxation time. (*Bichler, K. J., Jakobi, B., Schneider, G. J., Dynamical Comparison of Different Polymer Architectures – Bottlebrush vs. Linear Polymer. Macromolecules 2021, 10.1021/acs.macromol.0c02104*)

results in 3.5 decades per measurement and therefore, several temperatures are needed. In the single temperature representation (Fig. 5.8a) the susceptibilities do not show pronounced features, which would relate to relaxation times. By applying the frequency temperature superposition (FTS) principle the frequency window can be significantly extended by shifting the susceptibilities to one reference temperature [8]. As seen in Fig. 5.8b the gained frequency range has now ~9 orders of magnitude.

Immediately, a peak and additionally, a shoulder is visible. This slight shoulder in the region of $10^{-3} < \omega\tau_s < 10^{-1}$ comes along with a power law of $\omega^{0.5}$ at the high frequency side and goes over into the power low of $\omega^{0.75}$ at low frequencies. This behavior is typical for PDMS, however, rather unusual for other polymers [34, 35]. The origin lays in the intermolecular relaxation. For PDMS, the intermolecular contributions are dominating, starting at $\omega\tau_s \approx 10^{-1}$. In the region of the relaxation peak, $\omega\tau_s \approx 1$ both contributions, inter- and intramolecular, have basically the same amplitude. Therefore, the shoulder can be seen as a crossover between inter- and intramolecular relaxation [35].

The region of $\omega\tau_s < 10^{-3}$ describes the polymer dynamics with the power law $\omega^{0.75}$ fulfilling the theoretical expectation for linear PDMS. In contrast, ω^1 would describe a Debye behavior as known from simple liquids [36]. The next power law, $\omega^{0.5}$ agrees with the power law of high molecular weight linear PDMS, as found by Herrmann et al. [37]. In this region the literature assumes an interplay of glassy and polymer dynamics. In the dielectric spectroscopy experiments, such a shoulder is missing. The reason behind that is that PDMS has only a perpendicular dipole moment, resulting in the α-relaxation. In order to detect polymer dynamics, known

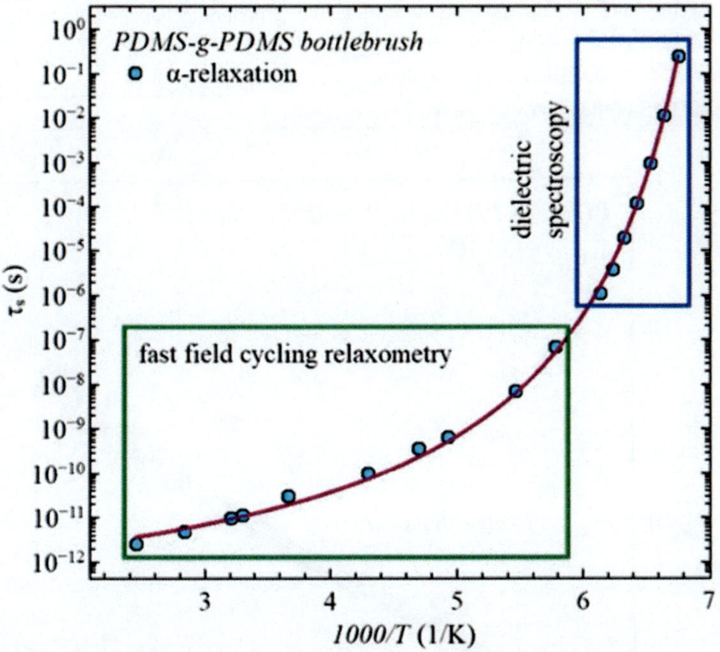

Fig. 5.9 Segmental relaxation time, τ_s, vs. 1000/T of the spherical PDMS-g-PDMS bottlebrush polymers measured with dielectric spectroscopy and fast field cycling relaxometry. Solid line is the best description with the VFT equation. (*Bichler, K. J., Jakobi, B., Schneider, G. J., Dynamical Comparison of Different Polymer Architectures – Bottlebrush vs. Linear Polymer. Macromolecules 2021, 10.1021/acs.macromol.0c02104*)

as normal mode in dielectric spectroscopy, a parallel dipole moment is required. As seen in Fig. 5.8b, the orange dashed line represents the relaxation peak at $T = -125$ °C obtained from dielectric spectroscopy normalized to one and overlaps in the peak region with the susceptibility master curve.

The relaxation peak, obtained by FFC relaxometry can be described with the HN-equation, while keeping the β-parameter (shape parameter for the left side) at 1 in order to exclude any effects from polymer dynamics. The resulting relaxation time is attributed to the α-relaxation of the reference temperature, T_{ref}, and can be related to the other measured temperatures by using the shift parameters, a_T. The obtained temperature dependent relaxation times can be combined with those, determined by dielectric spectroscopy in order to extend the time and temperature range (Fig. 5.9).

The results of both instruments can be described together with the VFT equation. With increasing temperature, the decay of the relaxation times slows down due to a saturation effect, occurring at infinite temperatures. The resulting fit parameters of the VFT description are summarized in Table 5.5.

Table 5.5 Resulting fit parameter for describing the temperature dependence of the relaxation times, determined by fast field cycling relaxometry and dielectric spectroscopy

	τ_∞ (s)	A (K)	T_0 (K)
PDMS-g-PDMS bottlebrush	$(6.26 \pm 0.1) \cdot 10^{-13}$	494 ± 16	130 ± 1

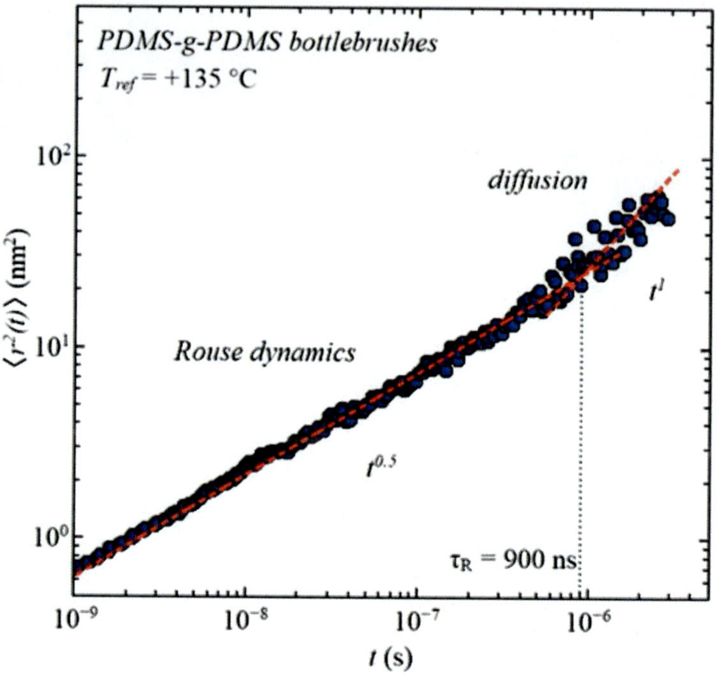

Fig. 5.10 Mean square displacement, $\langle r^2(t) \rangle$, vs. time, t, of the PDMS-g-PDMS bottlebrush polymer at the reference temperature $T_{\mathrm{ref}} = +135$ °C. The red dashed lines represent the two different power laws expected from the Rouse model for segmental dynamics and center of mass diffusion. The crossover between these two are known as the Rouse relaxation time, τ_R. (*Bichler, K. J., Jakobi, B., Schneider, G. J., Dynamical Comparison of Different Polymer Architectures – Bottlebrush vs. Linear Polymer. Macromolecules 2021, 10.1021/acs.macromol.0c02104*)

Taking the cosine transformation of the susceptibility master curve, in the relaxation rate representation, $R_1(\omega)$ vs. ω, according to Eq. (2.53), results in the mean square displacement $\langle r^2(t) \rangle$ as shown in Fig. 5.10.

Here two regions are obvious. Starting at short times, a power low of $t^{0.5}$ is present, whereby after $t = 900$ ns the second regime starts. This can be described by a power law of t^1. The Rouse model yields two power laws. At the short times, segmental relaxation results in $\langle r^2(t) \rangle \propto t^{0.5}$ whereas at longer times, the center of mass diffusion leads to $\langle r^2(t) \rangle \propto t^1$. These expectations are compatible with the experimental data. Furthermore, the intersection equals the longest relaxation time, the so-called Rouse time, τ_R [38]. In the case of the PDMS-g-PDMS bottlebrush polymer, this comes out as $\tau_R = 900$ ns (Fig. 5.10).

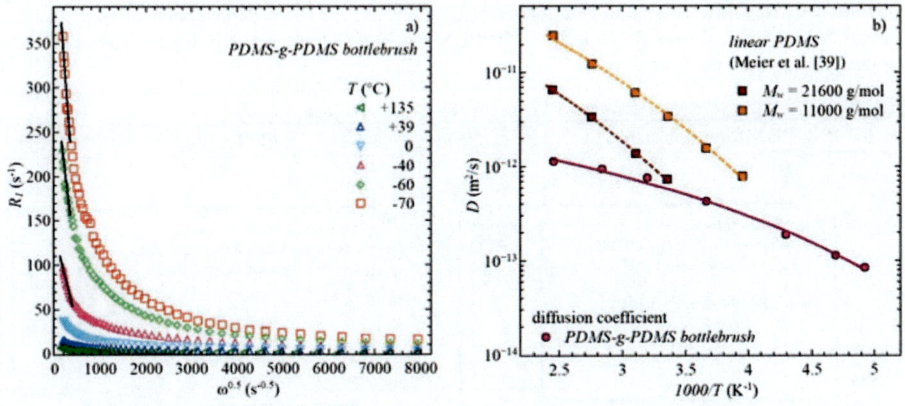

Fig. 5.11 (**a**) Relaxation rate, R_1, vs. $\omega^{0.5}$ of the PDMS-g-PDMS bottlebrush polymers in the temperature range $T = +135$ °C to $T = -70$ °C. Solid black lines represent the linear interpolation at the low frequencies. (**b**) Diffusion coefficient, D, vs. 1000/T for the spherical PDMS-g-PDMS bottlebrush polymers. Full purple line represents the best description with the VFT equation, Eq. (5.12). Dashed lines represent literature data for linear PDMS of different molecular weight M_w [39].

In the approximation of Rouse dynamics, determined by the mean square displacement, the diffusion coefficient based on Eq. (2.57) can be extracted by plotting R_1 vs. $\omega^{0.5}$ as shown in Fig. 5.11a. The diffusion coefficient results from the slope at the low frequency region, where the relaxation rate can be linearly interpolated.

The obtained temperature dependent diffusion coefficients, D, as a function of 1000/T are shown in Fig. 5.11b. Additionally, comparative literature values for linear PDMS with intermediate molecular weights, M_w, are also included [39].

While the diffusion coefficients of linear PDMS with different molecular weights are shifting almost parallel to each other, the bottlebrush polymer shows a reduced temperature dependence.

The diffusion coefficients of the bottlebrushes change by a factor of less than half a decade in a $\Delta T \sim 100$ °C increment. In contrast, the diffusion coefficient of the linear PDMS changes by a factor of around 10 in the same temperature interval. This suggests a reduced mobility in the segments once the diffusion regime is reached. The reason for that could lay in the restriction of the mobility due to interpenetrated side chains of neighbored bottlebrushes.

The resulting temperature dependence of the extracted segmental diffusion coefficients, D, shows a VFT behavior. This dependence can be described with a VFT equation similar to the temperature dependence of the relaxation times, τ_s, whereby here, D_∞ denotes the diffusion coefficient in the limit of high temperatures.

$$D = D_\infty \exp\left(\frac{B}{T - T_0}\right) \tag{5.12}$$

Table 5.6 Resulting fit parameter for describing the temperature dependence of the diffusion coefficient, extracted from R_1 vs. $\omega^{0.5}$ for the PDMS-g-PDMS bottlebrush polymer in a temperature range of $T = +135$ °C to $T = -70$ °C, with the VFT equation, Eq. (5.12)

	D_∞ (m^2/s)	B (K)	T_0 (K)
PDMS-g-PDMS bottlebrush	$(4.2 \pm 0.1) \cdot 10^{-12}$	(378 ± 84)	108 ± 17

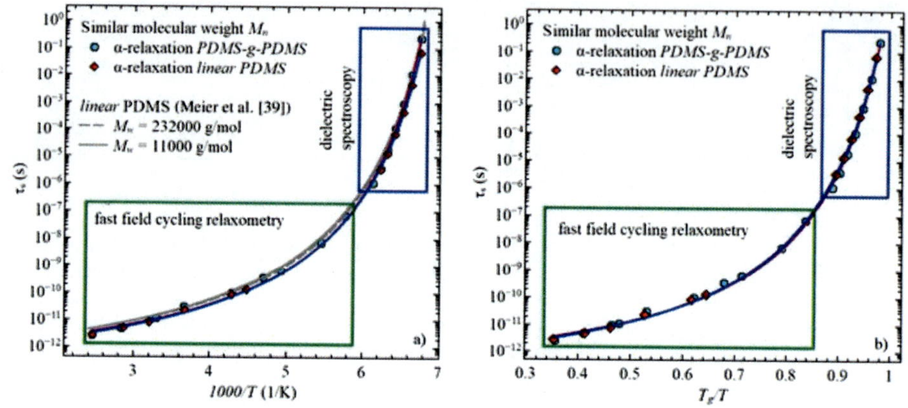

Fig. 5.12 (**a**) Segmental relaxation times, τ_s, vs. 1000/T for the PDMS-g-PDMS bottlebrush polymers and the single linear PDMS side chains of dielectric spectroscopy and fast field cycling relaxometry. Gray lines are literature values [39] and dark solid lines are the best description with VFT equation. (**b**) Segmental relaxation times, τ_s, vs. T_g/T for the PDMS-g-PDMS bottlebrush polymers and the single linear PDMS side chains of dielectric spectroscopy and fast field cycling relaxometry. Solid lines are the best description with the VFT equation

As seen in Fig. 5.11b, with increasing temperature, the diffusion coefficient increases as well. The obtained fit values, for the PDMS-g-PDMS bottlebrush polymer, by applying Eq. (5.12) are summarized in Table 5.6. Related fit values for the literature data can be found in the respective publication [39].

A comparison of the segmental relaxation times, τ_s, for the PDMS-g-PDMS bottlebrush and the linear PDMS polymer of similar molecular weight (Table 5.2) is shown in Fig. 5.12a. Hereby, a combination of dielectric spectroscopy and fast field cycling relaxometry was used, in order to extend the available time range.

Little deviations are visible in the dielectric region (low temperature) which can be explained by the slightly different glass transition temperature of PDMS-g-PDMS bottlebrush polymer compared to the linear PDMS. The reduced representation, related to the respective T_g as seen in Fig. 5.12b shows a full overlap of the two samples over the entire temperature range. Furthermore, literature data of linear PDMS with different M_w also overlap in the same region as the experimental values (Fig. 5.12a) [39]. This supports the independence of the α-relaxation from the molecular weight and one could assume, an independence on the polymer architecture. The gap between the data sets of FFC and dielectric spectroscopy, especially in the high molecular weight linear PDMS is caused by the fact, that high molecular

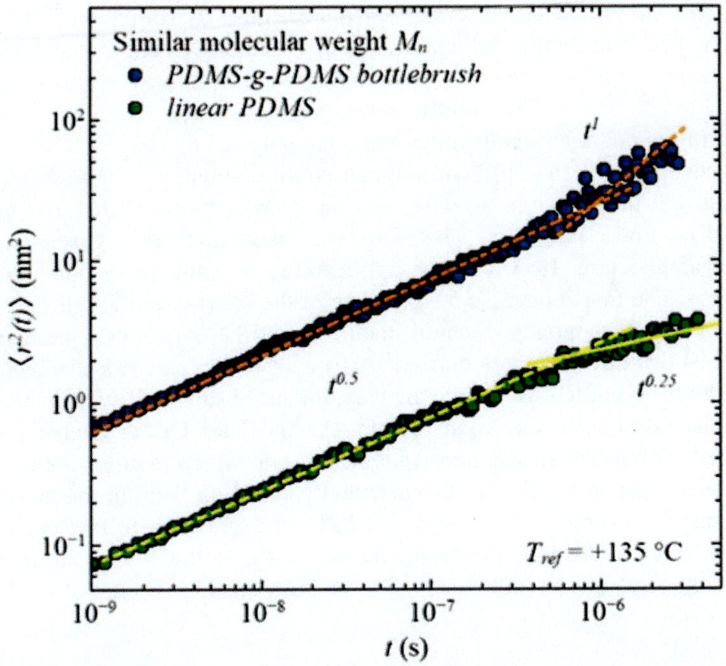

Fig. 5.13 Mean square displacement, $r^2(t)$, vs. time, t, of PDMS-g-PDMS bottlebrush polymer and linear PDMS of similar molecular weight at $T_{\mathrm{ref}} = +135$ °C. Dashed lines are the evolving power laws, associated with different regimes of the polymer dynamics. The green curve is shifted by a factor of 10^1 to lower values for the purpose of clearness. (*Bichler, K. J., Jakobi, B., Schneider, G. J., Dynamical Comparison of Different Polymer Architectures – Bottlebrush vs. Linear Polymer. Macromolecules 2021, 10.1021/acs.macromol.0c02104*)

weight PDMS tends to crystallizes very fast, which makes several temperatures impossible to measure [40].

In a next step the dynamics at the larger length scales of these two polymers are investigated. Therefore, the mean square displacement, obtained by FFC of both samples is compared and displayed as a function of time, t, in Fig. 5.13. Here $r^2(t)$ of the linear PDMS is shifted by a factor of 10^1 to lower values for the sake of clarity.

In both polymers, two different power law regimes develop. At short times, both polymers show the same power law of $t^{0.5}$, suggesting Rouse dynamics. While the bottlebrush continues with a power law of t^1 known for diffusion, starting at $\tau_R = 900$ ns, the linear PDMS shows a power law of $t^{0.25}$ starting at $t = 600$ ns. This regime denotes the constraint rouse regime. Continuing to longer times, would result in reputation, $r^2(t) \propto t^{0.5}$, and finally diffusion, $r^2(t) \propto t^1$. For FFC these two regions are beyond the detection window.

The regime of constraint Rouse dynamics is only accomplished by polymers with high molecular weight, where entanglement effects are present. This comparison shows that even if the PDMS-g-PDMS bottlebrush polymer has a high molecular weight, the dynamics pursue still the dynamics of low molecular weight linear

PDMS with no entanglement effect and pure Rouse dynamics until the onset of diffusion. This is accompanied with a shift of the entanglement molecular weight to higher values, simply by attaching side chains to a linear backbone, as suggested by rheology measurements [41]. Furthermore, the segmental relaxation is almost independent of the polymer architecture, while the polymer dynamics show remarkable differences related to the different polymer conformations.

So far, the mean square displacement analysis of the FFC data has only been explored on linear polymers. Therefore, the theory behind is based on a linear polymer architecture. However, the applicability to bottlebrush polymers can be questioned. The first concern is the meaning of the intermolecular relaxation, which is the important component for information about the dynamics. One could argue that at least two different intercontributions are available, e.g., side chain–side chain and bottlebrush–bottlebrush. Also, the way, the mean square displacement has been derived is based on a Gaussian system. As the later QENS analysis shows, a bottlebrush polymer melt is a non-Gaussian system which requires a non-Gaussian parameter for the mean square displacement calculation. Within the dissertation it was planned to combine the analysis of FFC and QENS to follow up with these concerns, but due to an unforeseen unavailability of the FFC instrument, these experiments were not possible and had to be postponed.

5.2.2 *Influence of the Shape Transition on the Dynamics*

Dielectric spectroscopy and quasi-elastic neutron scattering have been performed on three different PDMS-g-PDMS bottlebrush polymers, already introduced in Sect. 4.2.3, discussing the morphology in good solvent condition. For the sake of convenience, all bottlebrushes have a different side chain length, $M_n = 298$, 1800 and, 11, 500 g/mol, but virtually the same backbone length, M_n~16, 500 g/mol. This leads to shapes ranging from more spherical to elongated.

5.2.2.1 Dielectric Spectroscopy of PDMS-g-PDMS Bottlebrush Polymers: Shape Transition

All samples, together with the single side chains, i.e., PDMS-g-PDMS bottlebrush and linear PDMS, have been measured with dielectric spectroscopy to get information about the segmental relaxation times, τ_s. The relaxation spectra, dielectric permittivity, ϵ'', vs. frequency, *f*, are included in the Appendix. Only the samples with short and middle side chain length show major differences compared to the previous samples. In these cases, the segmental relaxation, influenced by the cold crystallization, the α_c-relaxation, is missing, due to the absence of crystallization, resulting from the low molecular weight of the side chains. In general, for all three samples, the detected α-relaxation is still governed by the side chains, since the majority of the dipole moments are located there.

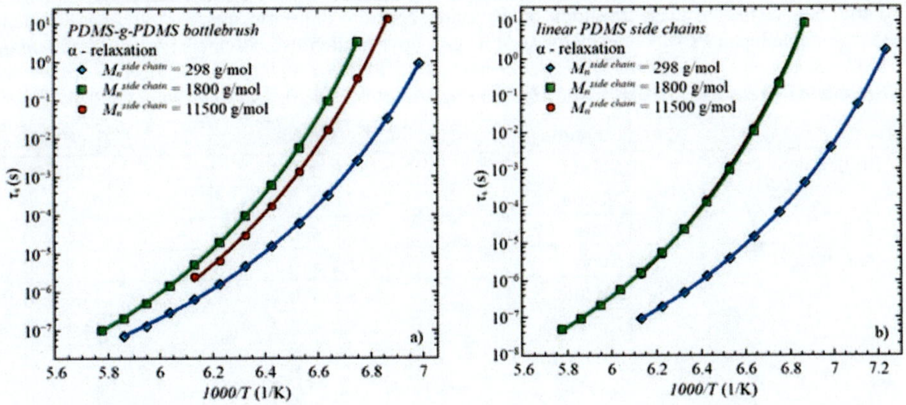

Fig. 5.14 Segmental relaxation times, τ_s, vs. 1000/T. (**a**) Each PDMS-g-PDMS bottlebrush polymer. (**b**) Each linear PDMS side chain. Solid lines are the best descriptions with the VFT equation

Table 5.7 Fit parameter for describing the temperature dependence of the segmental relaxation times from PDMS-g-PDMS bottlebrush polymer and their respective single side chain (*Jakobi, B., Bichler, K. J., Sokolova, A., Schneider, G. J., Dynamics of PDMS-g-PDMS Bottlebrush Polymers by Broadband Dielectric Spectroscopy. Macromolecules 2020, 10.1021/acs.macromol.0c01277*)

	Sample	τ_∞ (s)	A (K)	T_0 (K)
Bottlebrush	Short	$(12.3 \pm 2.3) \cdot 10^{-13}$	508 ± 11	124.5 ± 0.3
	Middle	$(13.4 \pm 3.9) \cdot 10^{-14}$	602 ± 17	128.6 ± 0.4
	Long	$(4.6 \pm 2.3) \cdot 10^{-14}$	654 ± 23	126.0 ± 0.4
Side chain	Short	$(23.1 \pm 1.8) \cdot 10^{-14}$	572 ± 4	118.8 ± 0.1
	Middle	$(92.3 \pm 6.1) \cdot 10^{-15}$	608 ± 4	126.7 ± 0.1
	Long	$(5.6 \pm 1.3) \cdot 10^{-15}$	759 ± 12	123.9 ± 0.2

Within this chapter, only the extracted temperature dependent relaxation times, i.e., τ_s vs. 1000/T, from the dielectric spectroscopy spectra will be included. Herefore, the normalized relaxation spectra of the α-relaxation have been used. Together with the shift factors of the x-direction, the temperature specific relaxation times are obtained. Associated fit parameters for the description of the dielectric spectroscopy spectra are summarized in the Appendix. The most interesting point, based on dielectric spectroscopy is the segmental relaxation, i.e., the α-relaxation. Therefore, in case of the bottlebrush with the longest side chain length, the influenced segmental relaxation, the α_c-relaxation is not taken into account for the data discussion, but the associated data, are included in the Appendix.

The relaxation times, τ_s, obtained for each PDMS-g-PDMS bottlebrush polymer together with the results for each respective linear PDMS side chain are shown in Fig. 5.14 as a function of 1000/T. Related fit parameters for the description with the empirical VFT equation are collected in Table 5.7.

Table 5.8 Extracted glass transition temperature, T_g, from dielectric spectroscopy measurement for the single linear PDMS side chains and the respective bottlebrush polymers (*Jakobi, B., Bichler, K. J., Sokolova, A., Schneider, G. J., Dynamics of PDMS-g-PDMS Bottlebrush Polymers by Broadband Dielectric Spectroscopy. Macromolecules 2020, 10.1021/acs.macromol.0c01277*)

	Sample	T_g (°C)	T_g (K)
Bottlebrush	Short	−132.8 ± 0.2	140.4 ± 0.2
	Middle	−127.0 ± 0.5	146.2 ± 0.5
	Long	−128.6 ± 0.7	144.6 ± 0.7
Side chains	Short	−137.3 ± 0.1	135.8 ± 0.1
	Middle	−128.8 ± 0.1	144.3 ± 0.1
	Long	−128.9 ± 0.2	144.3 ± 0.2

The temperature dependence shows strong similarities to the previously reported ones. With increasing temperature, the relaxation time speeds up. Inspecting the relaxation times for the bottlebrush polymers, as shown in Fig. 5.14a, it is obvious that the segmental relaxation times of all three samples differ from each other. Intuitively one would think that the order of the relaxation times follows the molecular weight from low to high. This is apparently not the case. The fastest bottlebrush has the shortest side chains, but the slowest one has the side chains with the middle molecular weight of $M_n = 1800$ g/mol.

The bottlebrush polymer having the longest side chains shows relaxation times which are in between of the bottlebrushes with short and medium side chain lengths. Comparing this behavior with the relaxation times of the single linear PDMS side chains shows different results (Fig. 5.14b). Instead of having the same order as the bottlebrush polymers, the relaxation times of the two linear PDMS serving as the long and middle side chains, superimpose and show identical relaxation times. Only the sample with the shortest side chains is faster, close to a factor of 100.

Segmental relaxation is always connected to the glass transition temperature, which can be determined as the temperature, at which the relaxation time $\tau_s = 100$ s, i.e., $T_g = T(\tau_s = 100\text{ s})$ [1]. Extracted values for the single side chains, as well as for the respective bottlebrush polymers are summarized in Table 5.8.

For linear polymers, the more flexible the polymer segments are, the lower the glass transition temperature will be [25]. PDMS is one of the most flexible polymers, resulting in one of the lowest glass transition temperatures. That is connected with the free volume induced by the chain ends. Polymers with low molecular weight, i.e., short polymer chains, have more chain ends, resulting in a higher free volume and finally showing a lower glass transition temperature compared to higher molecular weights [42]. Increasing the molecular weight reduces the free volume. However, after a polymer specific threshold value of the molecular weight, M_n, further decrease of the free chain ends has no impact on the glass transition temperature anymore and T_g stays constant as seen in Fig. 5.15.

Additionally, the glass transition temperatures for the single PDMS linear side chains and the respective bottlebrush polymers are also included in Fig. 5.15. Here is it clear that the T_g of the side chains with $M_n = 1800$ g/mol and $M_n = 11,500$ g/mol

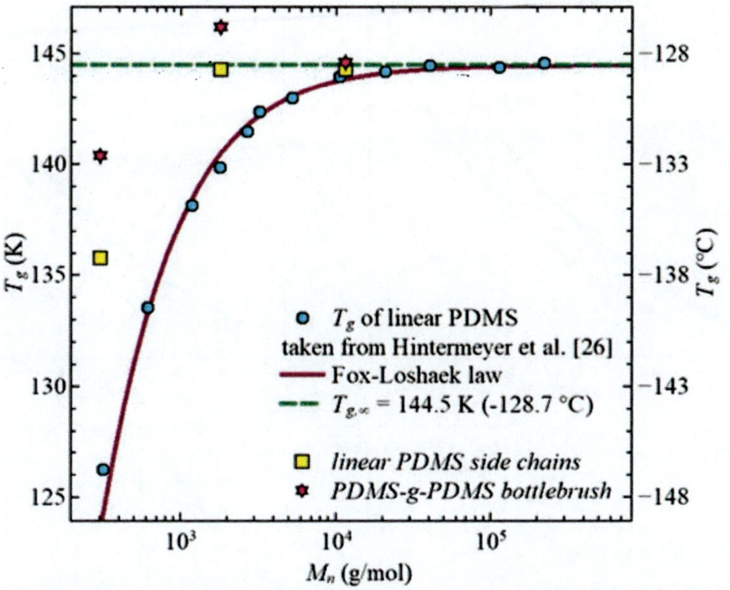

Fig. 5.15 Glass transition temperature, T_g, vs. number average molecular weight, M_n, of literature values, taken from Hintermeyer et al. [26] together with values for single linear PDMS side chains and the respective PDMS-g-PDMS bottlebrush polymers. Full purple line represents the description with the Fox–Loshaek law (c.f. Appendix equation (B.1)). Green dashed line represents the $T_{g,\infty}$ for infinite molecular weights

has the same value and seems to be already in the plateau region. This is the reason for the overlapping relaxation times in Fig. 5.14b. The value of the middle molecular weight $M_n = 1800$ g/mol shows a value contradicting the literature. Taking into account the different end groups compared to the literature (literature: two CH_3-groups, here: one C_4H_9-group and one OH-group) it gives rise to the possibility of a lowered glass transition temperature due to a bulkier end group which generates a higher free volume. Instead, the T_g is higher compared to literature and similar to the side chains of $M_n = 11,500$ g/mol. This might origin in the hydrogen bonding induced by the second end group. However, the deviations of the smallest side chains can be well explained with the different end groups (literature: two CH_3-groups, here: one SiH and one OH), generating less free volume due to the linear structure, compared to literature, which results in a higher glass transition temperature as normally expected.

The relaxation behavior of the bottlebrush polymers can also be explained with the different glass transition temperature as seen in Fig. 5.15. However, due to the different end groups of the sample with the smallest side chains compared to the other two samples, it is rather unpractical to compare all three together. Therefore, the sample having the shortest side chains is excluded.

Recalling the saturated T_g of the single side chains, one would expect, that also the T_g of the respective bottlebrush polymers is saturated. However, the grafting

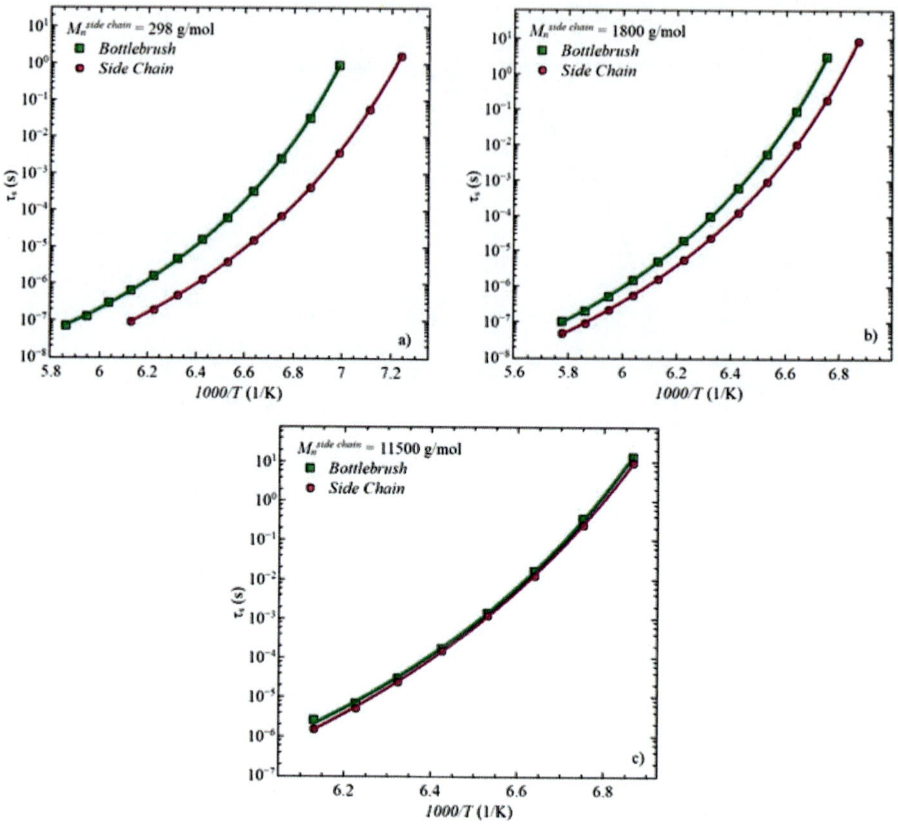

Fig. 5.16 Segmental relaxation time, τ_s, vs. 1000/*T* for the PDMS-g-PDMS bottlebrush polymer compared with their respective single linear PDMS side chains. (**a**) $M_n^{\text{side chain}} = 298$ g/mol, (**b**) $M_n^{\text{side chain}} = 1800$ g/mol, (**c**) $M_n^{\text{side chain}} = 11,500$ g/mol. Solid lines are the best description with the VFT equation. (*Jakobi, B., Bichler, K. J., Sokolova, A., Schneider, G. J., Dynamics of PDMS-g-PDMS Bottlebrush Polymers by Broadband Dielectric Spectroscopy. Macromolecules 2020, 10.1021/acs.macromol.0c01277*)

density slightly differs from these two samples (Table 4.8) which induces more free volume in case of the bottlebrush with long side chains. This could cause the lower glass transition temperatures in case of the long side chains. In order to fully confirm this assumption, further experiments with bottlebrush polymers, having a variety of grafting densities would be needed. In general, branched polymers, e.g., bottlebrush polymers, have a higher glass transition temperature as their linear unbranched counter parts [43]. Due to the different glass transition temperatures, the temperature dependent relaxation times show the particular behavior as seen in Fig. 5.14a.

A comparison of the relaxation times, τ_s, for each PDMS-g-PDMS bottlebrush polymer with their respective side chains is illustrated in Fig. 5.16 as a function of 1000/*T*. Here, the same behavior is indicated as discussed earlier, but a trend based on the side chain lengths is obvious. With increasing molecular weight of the side

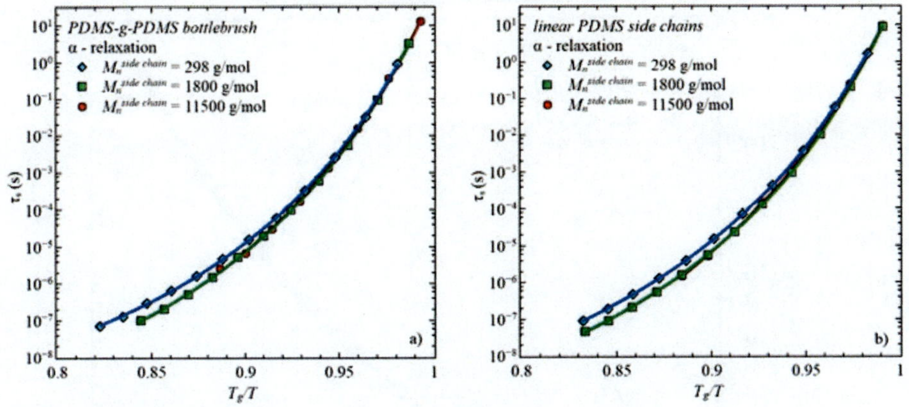

Fig. 5.17 Segmental relaxation times, τ_s, vs. T_g/T. (**a**) PDMS-g-PMDS bottlebrush polymer. (**b**) Single linear PDMS side chains. Solid lines are the best description with the VFT equation

chains, the slowdown process, associated with the grafting process weakens. While at the shortest side chain length (Fig. 5.16a) the strongest reduction of the relaxation times is found, ~2.5 decades, this reduces to ~1.5 decades in case of the middle side chains length (Fig. 5.16b). Upon increasing the side chain length further, to $M_n^{\text{side chain}} = 11,500$ g/mol, both samples, bottlebrush and single side chain, coincide over the entire temperature range and no reduction of the relaxation time is visible (Fig. 5.16c).

For $M_n^{\text{side chain}} = 298$ g/mol, the side chain consists of four segments. Therefore, attaching one segment to the flexible PDMS backbone has the greatest impact, since the other three chain segments are also impacted, which results in the reduced relaxation times. However, increasing the molecular weight of the side chains is connected with increasing the distance from the branch point to the outer side chain segment. This reduces the influences from the attached chain segment on the whole side chain. Therefore, the reduction in relaxation times is not as pronounced as for the shortest side chain length, due to the averaging with less influenced segments. For higher molecular weights, $M_n^{\text{side chain}} = 11,500$ g/mol the whole side chain does not experience any changes caused by fixing one end to the backbone. Therefore, the segmental relaxation times of bottlebrush and single linear side chain superimpose.

Based on this comparison, a threshold molecular weight of around $M_n^{\text{side chain}}\tilde{1}0,000$ g/mol can be assumed to separate the side chain length in bottlebrushes, below which the segmental relaxation time is influenced by the architecture.

Normalizing the relaxation times of the three samples to their respective glass transition temperatures, deviations independent of the different glass transition temperatures become visible. As seen in Fig. 5.17a, at the low temperatures, the relaxation times of all three PDMS-g-PDMS bottlebrush polymers overlap, but at higher temperatures, only the bottlebrush polymers with long and middle side chain length coincide. The bottlebrush polymer with the shortest side chains shows longer

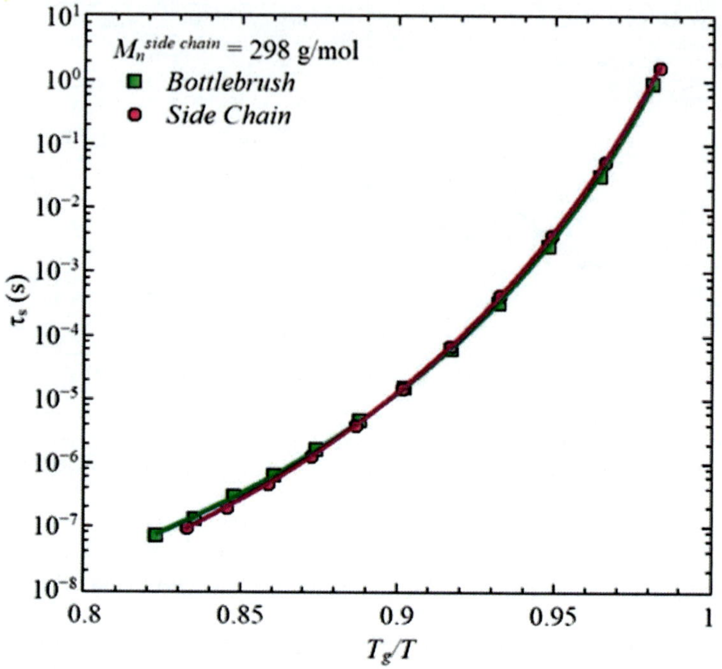

Fig. 5.18 Segmental relaxation times, τ_s, vs. T_g/T for the PDMS-g-PDMS bottlebrush polymer and their respective single side chains for the sample having the shortest side chains, $M_n^{\text{side chain}} = 298$ g/mol. Solid lines are the best description with the VFT equation

relaxation times at higher temperatures compared to the other two samples. In this sample, the backbone has a significant contribution to the overall molecular weight.

This might lead to the assumption, that at higher temperatures, the backbone significantly contributes to the segmental relaxation, slowing the average α-relaxation down, due to the inertia of the backbone segments themselves. However, looking at the pure linear PDMS side chains (Fig. 5.17b), the same effect is observed. This contradicts the previous assumption. Therefore, it can be assumed that the difference originates either from the different end groups or from the extremely low molecular weight.

Comparing the bottlebrush polymer with the respective side chains, normalized to the glass transition temperature, shows very good overlap across the whole temperature range. This is exemplified for the bottlebrush polymer having the shortest side chains in Fig. 5.18.

In summary the relaxation behavior of bottlebrush polymers is still dominated by the side chain contribution, even for polymers with a significant backbone fraction. Furthermore, it appears that the grafting density has great impact on the relaxation behavior of bottlebrush polymers. The extracted glass transition temperatures seem to be affected by the number of side chains. Independently of the conformation, end groups influence the relaxation behavior as well. However, for a more detailed analysis about the end groups, a more detailed study based on polymers with different end groups would be needed.

5.2.2.2 Quasi-Elastic Neutron Scattering (QENS) on PDMS-g-PDMS Bottlebrush Polymers: Shape Transition

In the next step, Quasi-Elastic Neutron Scattering (QENS) was conducted to obtain more information on the different relaxations in these samples. As described earlier, three different instruments have been used to cover a time range of $t = 1$ ps to $t = 1$ ns, which includes the segmental relaxation and faster dynamics, both with spatial resolution in the range of $Q = 0.6 - 1.8$ Å^{-1}. The data analysis for all three samples has been done in a similar way. The procedure is shown exemplary on the PDMS-g-PDMS bottlebrush with the longest side chains, $M_n^{\text{side chain}} = 11,500$ g/mol and then all three samples will be compared, while the analysis of the other two samples is shown in the Appendix.

After transforming the data into the time domain, a combination of all three instruments is possible as seen in Fig. 5.19, exemplary for three selected Q-values,

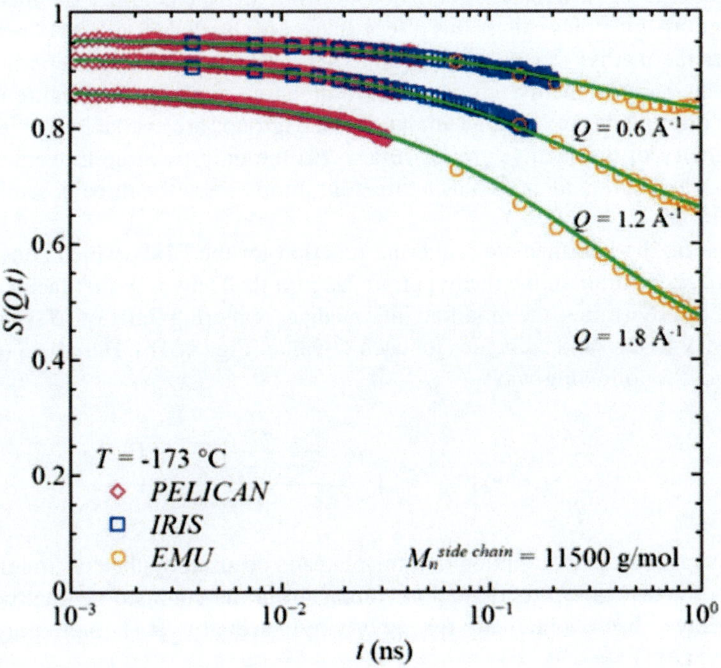

Fig. 5.19 Intermediate scattering function, $S(Q, t)$, vs. time, t, for the PDMS-g-PDMS bottlebrush polymer with $M_n^{\text{side chain}} = 11,500$ g/mol. The three colors indicate the three different instruments: Pink: PELICAN, Blue: IRIS, and Orange: EMU. The solid lines are the best descriptions of the data

$Q = 0.6$, 1.2, and, 1.8 $Å^{-1}$ at one specific temperature, $T = -173$ °C. The different colors indicate the three different instruments. Hereto, the data of IRIS and EMU were adjusted to merge with the PELICAN data. The time-of-flight instrument (PELICAN) measures the biggest energy transfers resulting in the highest accuracy regarding their scattering intensity.

From the EFW scan, the combined QENS time region captures two processes: (1) methyl group reorientation and (2) segmental dynamics. At low temperatures, $T = -198, -173, -148, -123$, and, -98 °C, only the methyl group reorientation is detectable. Here the segmental relaxation can be considered as frozen and is too slow to relax in the available time window [1]. At $T = -23$ °C and $T = +27$ °C, the methyl group rotation is too fast for the time window and the segmental relaxation determines the measurement signal.

At the lowest temperatures, a weak decay with almost Q-independent relaxation times is observed in the accessible time range, as seen in Fig. 5.20a, b. This indicates a localized motion, e.g., the reorientation of the methyl groups [9]. With increasing temperature, the amplitude decreases slightly and the decay is still almost independent of the momentum transfer, which adds further support for the localized motions.

In the long-time region, a plateau, intensifying with increasing temperature, emerges (Fig. 5.20c, d). The most obvious plateau is observed at the highest temperature, dominated exclusively by the methyl group dynamics, i.e., $T = -98$ °C, as seen in Fig. 5.21a. This raises the question whether the confined geometry of the methyl group dynamics causes this effect. In case of the PDMS-g-PDMS bottlebrush polymers the methyl groups are very dominant, since every monomeric unit of the side chains has two methyl groups. Due to the large number of repeating units and the high grafting density, a huge amount of CH_3-groups are available. Remembering the geometry of the methyl groups (Fig. 5.2a) the only possible movement of the methyl group can be identified as a threefold jump across the three possible hydrogen positions.

Adjusting the intermediate scattering function for the EISF, which considers the geometrical constrain of the methyl group, and for the Debye–Waller factor, describing very fast dynamics, the modified intermediate scattering function, $S'(Q,t)$, shows a full decay to zero and coincides for each Q-value (Fig. 5.21b). Hereby, $S'(Q,t)$ was derived in the following way:

$$S'(Q,t) = \exp\left(-\frac{t}{\tau_\ell}\right)^{\beta_\ell} = \frac{\frac{S(Q,t)}{A\cdot\exp\left(-\frac{1}{3}u^2\cdot Q^2\right)} - \mathrm{EISF}}{(1-\mathrm{EISF})} \tag{5.13}$$

This supports the assumption that the plateau is caused by the constraint dynamics of the methyl group reorientation coming from the confined geometry. Similar results have been obtained for poly(vinyl acetate) [44] and poly(methyl methacrylate) [13].

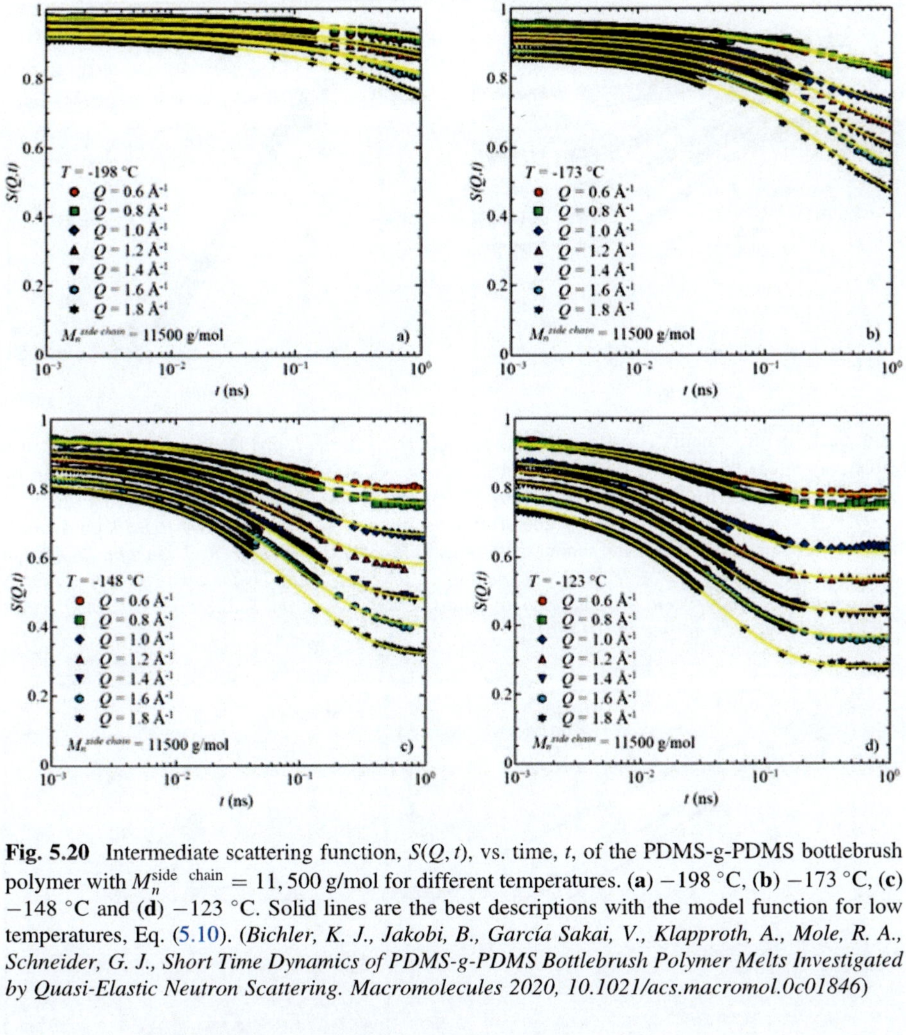

Fig. 5.20 Intermediate scattering function, $S(Q,t)$, vs. time, t, of the PDMS-g-PDMS bottlebrush polymer with $M_n^{\text{side chain}} = 11{,}500$ g/mol for different temperatures. (**a**) -198 °C, (**b**) -173 °C, (**c**) -148 °C and (**d**) -123 °C. Solid lines are the best descriptions with the model function for low temperatures, Eq. (5.10). (*Bichler, K. J., Jakobi, B., García Sakai, V., Klapproth, A., Mole, R. A., Schneider, G. J., Short Time Dynamics of PDMS-g-PDMS Bottlebrush Polymer Melts Investigated by Quasi-Elastic Neutron Scattering. Macromolecules 2020, 10.1021/acs.macromol.0c01846*)

5.2.2.2.1 Geometry of the Methyl Groups

Investigation of the Q-dependence of the EISF shows decreasing values with increasing momentum transfer and increasing temperature (Fig. 5.22a). Increasing momentum transfer is associated with a decreasing size of the observed object accessible to the experiment. Thus, with decreasing observation size, the restriction of the motion decreases and allows more freedom for the reorientation. Upon increasing temperature, the overall process accelerates and requires more space. This can be shown by analyzing the extracted EISF values based on the underlying potential of the threefold jump, Eq. (2.28), resulting in temperature dependent radii, R, representing the jumping distance. As seen in Fig. 5.22b, the radius, R, shows a tendency of saturation at high temperatures.

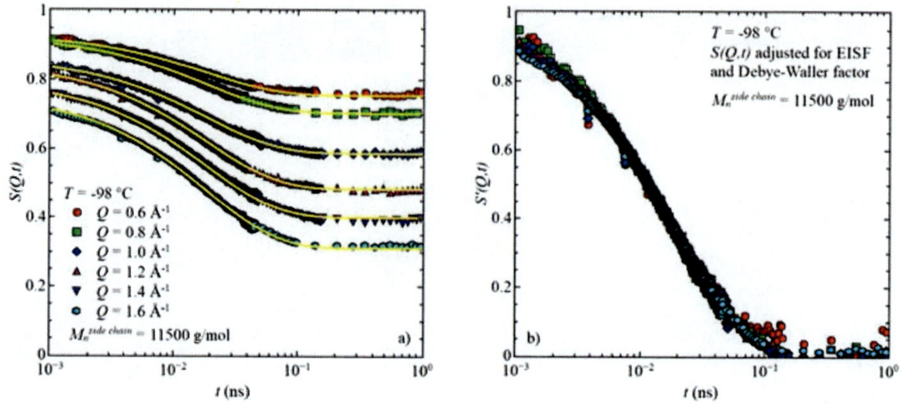

Fig. 5.21 (**a**) Intermediate scattering function, $S(Q, t)$, vs. time, t, of PDMS-g-PDMS bottlebrush polymer with $M_n^{\text{side chain}} = 11,500$ g/mol at $T = -98$ °C. Solid lines are the best description using the model function for low temperatures, Eq. (5.10). (**b**) Adjusted intermediate scattering function, $S'(Q, t)$, for EISF and the Debye–Waller factor vs. time, t, for the PDMS-g-PDMS bottlebrush polymer at $T = -98$ °C for the same Q-values as in Fig. 5.21a. (*Bichler, K. J., Jakobi, B., García Sakai, V., Klapproth, A., Mole, R. A., Schneider, G. J., Short Time Dynamics of PDMS-g-PDMS Bottlebrush Polymer Melts Investigated by Quasi-Elastic Neutron Scattering. Macromolecules 2020, 10.1021/acs.macromol.0c01846*)

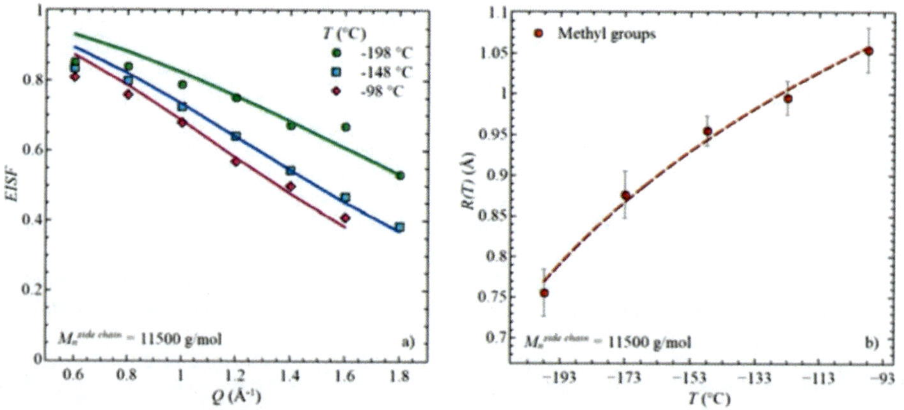

Fig. 5.22 (**a**) Elastic incoherent structure factor, EISF, vs. momentum transfer, Q, for three selected temperatures, $T = -198$ °C, $T = -148$ °C and, $T = -98$ °C for PDMS-g-PDMS bottlebrush polymer with the longest side chains. Solid lines represent the data description with the threefold jump model (Eq. 2.28). Errors are within symbol size. (**b**) Radius, R, vs. temperature, T, for the PDMS-g-PDMS bottlebrush polymer with the longest side chains. Dashed line is guide for the eye. (*Figure (**a**) taken from Bichler, K. J., Jakobi, B., García Sakai, V., Klapproth, A., Mole, R. A., Schneider, G. J., Short Time Dynamics of PDMS-g-PDMS Bottlebrush Polymer Melts Investigated by Quasi-Elastic Neutron Scattering. Macromolecules 2020, 10.1021/acs.macromol.0c01846*)

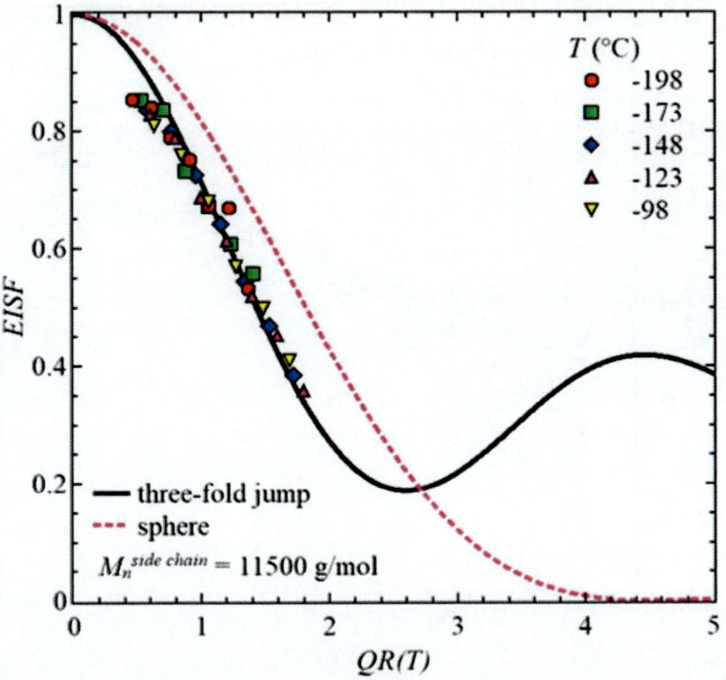

Fig. 5.23 Elastic incoherent structure factor, EISF, vs. $QR(T)$ of PDMS-g-PDMS bottlebrush polymers with $M_n^{\text{side chain}} = 11{,}500$ g/mol for different temperature. Solid black line represents the threefold jump model, dashed pink line represents the potential for localized motion within a sphere. (*Bichler, K. J., Jakobi, B., García Sakai, V., Klapproth, A., Mole, R. A., Schneider, G. J., Short Time Dynamics of PDMS-g-PDMS Bottlebrush Polymer Melts Investigated by Quasi-Elastic Neutron Scattering. Macromolecules 2020, 10.1021/acs.macromol.0c01846*)

This seems plausible, because with increasing temperature, the mobility of the polymer as well as of the methyl groups increases. With increasing mobility, more space is required and upon reaching the final extension of the methyl groups the radius saturates. The determination of a saturation radius cannot be performed due to a missing theoretical description of the temperature dependence of the threefold jump and the small temperature range over which it is observed.

Removing the temperature dependence of the EISF, by multiplying the momentum transfer, Q, with the radius for each temperature, R, further supports the use of the threefold jump model for the methyl group reorientation. As seen in Fig. 5.23, the adjusted experimental data can be well described by the EISF for the threefold jump. Additionally, the theoretical potential for motions within a sphere is illustrated, but cannot describe the data. This is in good agreement with available literature about methyl group dynamics [12–14].

A comparison of all three samples regarding their temperature dependent radii, R (T), shows in general similar trends with the temperature (Fig. 5.24). With increasing temperature, the radius increases in all cases. However, while for the sample with the

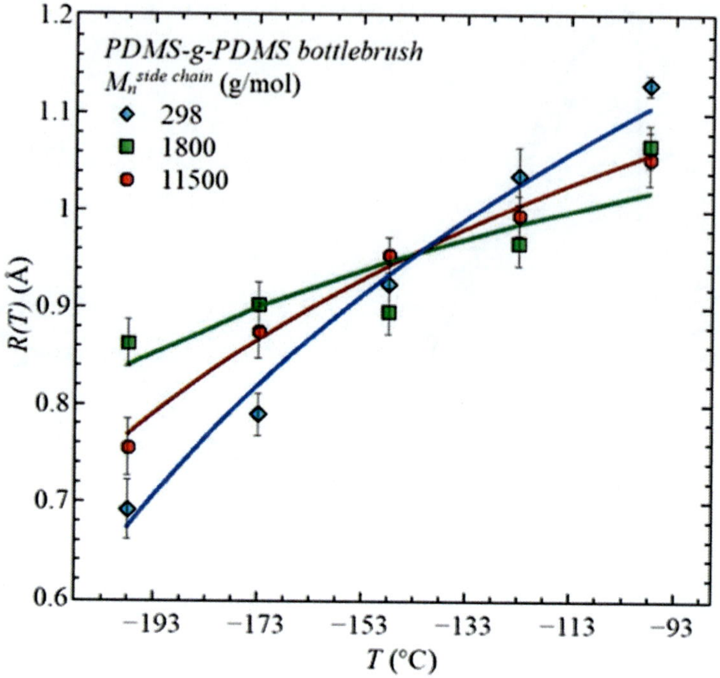

Fig. 5.24 Radius, *R*, vs. temperature, *T*, for the three different PDMS-g-PDMS bottlebrush polymers as indicated. Solid lines are guide for the eye

longest side chains, the radius changes by around $\Delta R = 0.30$ Å, the radius of the sample with the shortest side chains increases by $\Delta R = 0.44$ Å within the same temperature change of $\Delta T = 100$ °C. In case of the sample with medium side chains, the radius has the weakest temperature dependence. Furthermore, all radii seem to have the same size of $R \sim 0.96$ Å at one specific temperature, $T \sim -143$ °C. For all the mentioned findings above, a quantitative discussion is not possible since a related theory based on the temperature dependence of the methyl groups, based on the radius, *R*, is not available.

5.2.2.2.2 Relaxation Times of Methyl Group Dynamics

Additionally, the relaxation times of the methyl groups were determined. As shown in Fig. 5.25, the relaxation times are almost independent of the side chain lengths and decreases with increasing temperature.

This relaxation can be well described with the Arrhenius law, Eq. (2.11), which indicates a thermally activated process, typical for the methyl group reorientation. The resulting fitting parameters for all three samples are summarized in Table 5.9.

Here, the activation energy, E_A, determines the amount of energy needed to perform one jump to the next hydrogen position (Fig. 5.2a). The obtained values

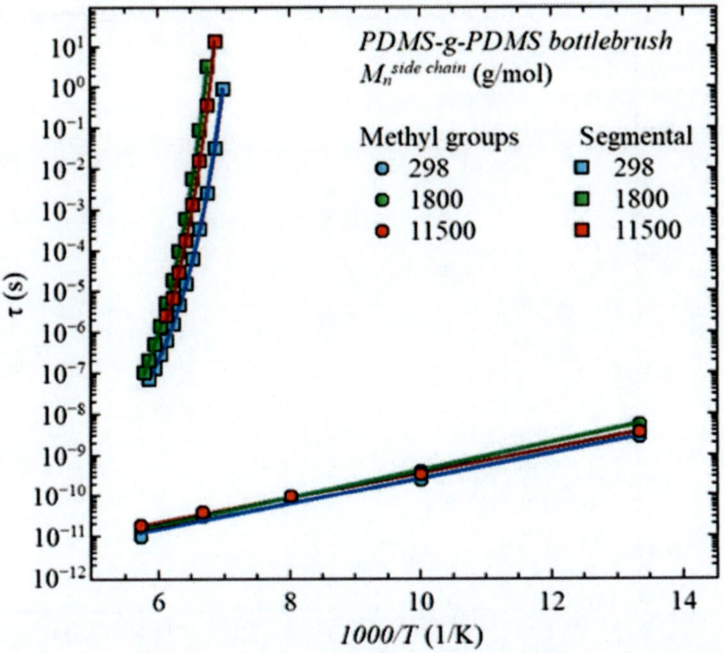

Fig. 5.25 Relaxation times, τ, vs. 1000/T. for the PDMS-g-PDMS bottlebrush polymers for the segmental relaxation (dielectric spectroscopy) and for the methyl group dynamics (QENS). The solid lines are data descriptions with the VFT equation (segmental relaxation) and the Arrhenius equation (methyl groups dynamics). Error bars are within symbol size and omitted. (*Bichler, K. J., Jakobi, B., García Sakai, V., Klapproth, A., Mole, R. A., Schneider, G. J., Short Time Dynamics of PDMS-g-PDMS Bottlebrush Polymer Melts Investigated by Quasi-Elastic Neutron Scattering. Macromolecules 2020, 10.1021/acs.macromol.0c01846*)

Table 5.9 Fitting parameter of the data description with the Arrhenius equation for the methyl group dynamics for the three samples with side chain molecular weight as indicated (*Bichler, K. J., Jakobi, B., García Sakai, V., Klapproth, A., Mole, R. A., Schneider, G. J., Short Time Dynamics of PDMS-g-PDMS Bottlebrush Polymer Melts Investigated by Quasi-Elastic Neutron Scattering. Macromolecules 2020, 10.1021/acs.macromol.0c01846*)

$M_n^{\text{side chain}}$ (g/mol)	τ_∞ (s)	E_A (J/mol)
298	$(2.04 \pm 0.72) \cdot 10^{-13}$	5969 ± 316
1800	$(1.67 \pm 0.11) \cdot 10^{-13}$	6522 ± 62
11, 500	$(3.21 \pm 0.22) \cdot 10^{-13}$	5844 ± 44

for all three samples are in the same order of magnitude. Literature values of QENS experiments on linear PDMS list an activation energy of $E_A = (4.5 \pm 0.5)$ kJ/mol [11, 12] or $E_A = 5.2$ kJ/mol [45].

In polymers, the methyl group dynamics are influenced by the chemical environment. Therefore, the activation energy is different for different polymers and it can be assumed that also a dependence on the polymer architecture is present. The

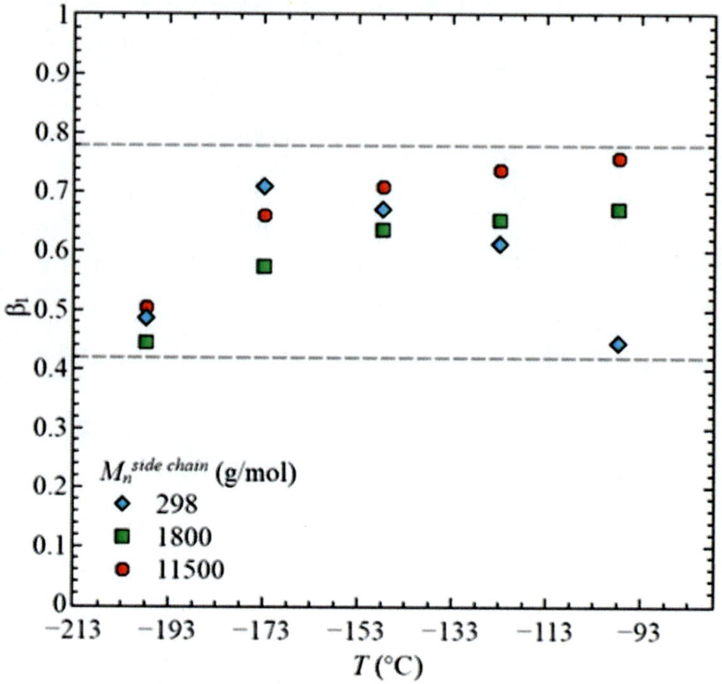

Fig. 5.26 Stretching parameter, β_ℓ, vs. temperature, T, of all three bottlebrush samples. Errors are within symbol size. (*Bichler, K. J., Jakobi, B., García Sakai, V., Klapproth, A., Mole, R. A., Schneider, G. J., Short Time Dynamics of PDMS-g-PDMS Bottlebrush Polymer Melts Investigated by Quasi-Elastic Neutron Scattering. Macromolecules 2020, 10.1021/acs.macromol.0c01846*)

standard deviation of the activation energy of the sample with the smallest side chains, $M_n^{\text{side chain}} = 298$ g/mol, is ~fivefold increased compared to the other two. This could originate from a higher dispersity in surroundings from the existing methyl groups, i.e., the backbone methyl groups have slightly different contributions.

Furthermore, also the relaxation times are affected by the different chemical environment. This results in a multitude of different relaxation times. To accomplish this, the Kohlrausch–Williams–Watts (KWW) function was used within the model function for the low temperatures. Hereby, the relaxation times are connected with a stretching parameter, β_ℓ, which is taken to be independent of the momentum transfer. All values for β_ℓ show a temperature dependence and have values between $\beta_{\ell_{\max}} = 0.78$ and $\beta_{\ell_{\min}} = 0.42$ (Fig. 5.26).

In case of the shortest side chains, β_ℓ fluctuates around β_ℓ~0.6. For the sample with the two longer side chain lengths, the stretching parameter increases with increasing temperature. However, at this point, it is hard to predict any relationship between stretching parameter, β_ℓ, and the temperature, T, or between β_ℓ and $M_n^{\text{side chain}}$. At least the latter two samples ($M_n^{\text{side chain}} = 1800, \ 11,500$ g/mol)

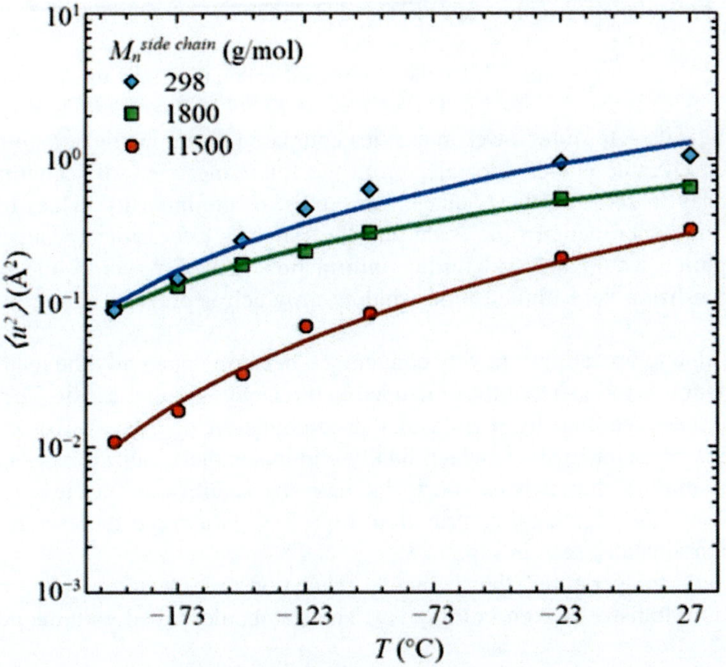

Fig. 5.27 Atomistic mean square displacement, $\langle u^2 \rangle$, vs. temperature, T, for all three samples, determined by the Debye–Waller factor. Solid lines are guidelines for the eyes. (*Bichler, K. J., Jakobi, B., García Sakai, V., Klapproth, A., Mole, R. A., Schneider, G. J., Universality of the Time-Temperature Scaling Observed by Neutron Spectroscopy on Bottlebrush Polymers. Nano Letters 2021, 10.1021/acs.nanolett.1c01379*)

seem to reach a constant value at high temperatures, whereas for the smallest sample, no systematic dependency is visible. These values seem to have a statistical distance from the average position, though a real explanation would require having a statistically relevant number of values.

general, the analysis of QENS data is conducted under the assumption that all relaxations which are too fast for the available time window can be captured by a single Debye–Waller factor, including the atomistic mean square displacement, $\langle u^2 \rangle$. Looking at the temperature evolution of $\langle u^2 \rangle$ as shown in Fig. 5.27 an increase with increasing temperature can be observed.

This is similar for all samples, whereby the absolute values depend systematically on the molecular weight of the side chains. Hereby, the shortest side chains show the largest atomic mean square displacement. Shorter chains are stiffer compared to longer chains and therefore, their frequency of vibration is much higher, resulting in larger mean square displacements of the atoms.

5.2.2.2.3 Relaxation Times of Segmental Relaxation

Increasing the temperature shifts the segmental relaxation into the accessible time window as seen for $T = -23$ °C and $T = +27$ °C in Fig. 5.28a, b, respectively. Here, the data sets show lower intensities compared to the methyl group dynamics and each Q-value is well separated from the following one. All scattering curves fully decay to zero for all Q-values. Since most of the intensity values reach zero within the experimental time scale and the range of accessible Q-values, higher temperatures would not add further information. In other words, the segmental relaxation has a very limited time window in which it prevails before it becomes zero.

The high temperatures are very convenient, because apparently the methyl group reorientation is too fast and thus included in the Debye–Waller factor. The assumption has been verified by a fit with the combination of Eqs. (5.10) and (5.11), yielding a constant EISF~1, which finally eliminates the localized motions. Therefore, to simplify the analysis and to increase the stability and accuracy of the fit values, only the segmental contribution, Eq. (5.11), was used to describe the high temperature data as seen in Fig. 5.28.

For both temperatures, the segmental relaxation time decreases with increasing momentum transfer as seen in Fig. 5.29. This can be described with the power law

$$\tau_s \propto Q^{-s} \tag{5.14}$$

whereby, in the Gaussian approximation, the exponent s can be connected to the stretching parameter β_s by $s = 2/\beta_s$ [9, 17, 18, 46]. Additionally, with increasing side

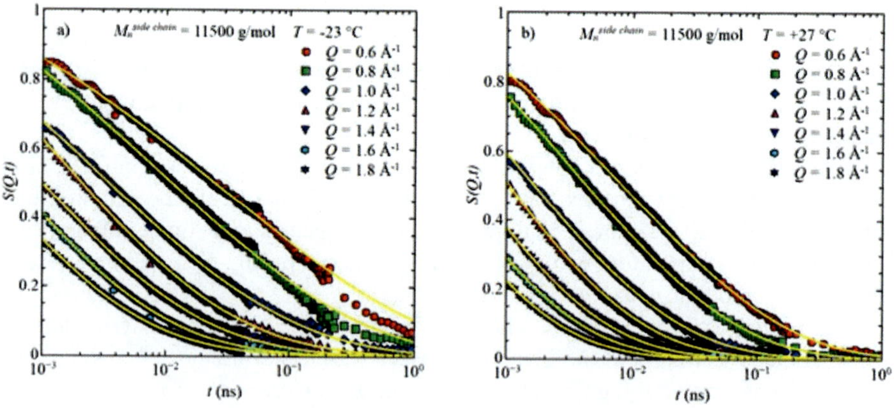

Fig. 5.28 Intermediate scattering function, $S(Q, t)$, vs. time, t, of the PDMS-g-PDMS bottlebrush polymer with the longest side chains for different temperatures. (**a**) $T = -23$ °C, and (**b**) $T = +27$ °C. Solid lines are the best description with the model function for high temperatures, Eq. (5.11). (*Bichler, K. J., Jakobi, B., García Sakai, V., Klapproth, A., Mole, R. A., Schneider, G. J., Short Time Dynamics of PDMS-g-PDMS Bottlebrush Polymer Melts Investigated by Quasi-Elastic Neutron Scattering. Macromolecules 2020, 10.1021/acs.macromol.0c01846*)

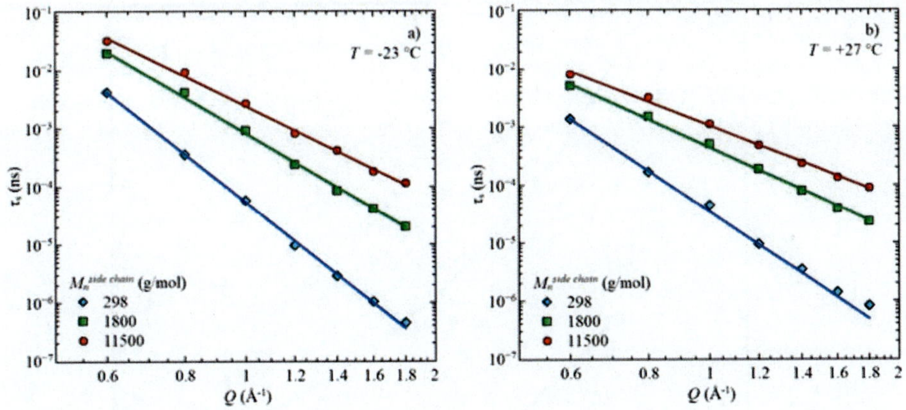

Fig. 5.29 Segmental relaxation time, τ_s, vs. momentum transfer, Q, for temperatures (**a**) $T = -23$ °C and (**b**) $T = +27$ °C for all three samples. (*Bichler, K. J., Jakobi, B., García Sakai, V., Klapproth, A., Mole, R. A., Schneider, G. J., Short Time Dynamics of PDMS-g-PDMS Bottlebrush Polymer Melts Investigated by Quasi-Elastic Neutron Scattering. Macromolecules 2020, 10.1021/acs. macromol.0c01846*)

Table 5.10 Fit parameter of all three samples for the power law of τ_s vs. Q (*Bichler, K. J., Jakobi, B., García Sakai, V., Klapproth, A., Mole, R. A., Schneider, G. J., Short Time Dynamics of PDMS-g-PDMS Bottlebrush Polymer Melts Investigated by Quasi-Elastic Neutron Scattering. Macromolecules 2020, 10.1021/acs.macromol.0c01846*)

Sample	s ($T = -23$ °C)	s ($T = +27$ °C)
Short	(8.52 ± 0.02)	(7.24 ± 0.09)
Middle	(6.35 ± 0.13)	(4.96 ± 0.12)
Long	(5.31 ± 0.12)	(4.24 ± 0.10)

chain length, the relaxation times increase. Also, a temperature effect is visible. Increasing the temperature, by $\Delta T = 50$ °C, accelerates the relaxation by around half a decade. This behavior is equal for each sample and the resulting power laws associated with each temperature and sample are summarized in Table 5.10.

For both temperatures a pronounced decrease of the slope with increasing side chain length is observed, as indicated by the s parameter. With increasing temperatures, s reduces, thus the Q-dependence of the relaxation time on the momentum transfer is reduced. The two samples having the longest side chains at $T = +27$ °C have a slope close to $s \sim 4$, which would suggest Gaussian behavior.[17, 18] However, the relationship between the parameter s, and the stretching parameter, β_s, within the Gaussian approximation ($s = 2/\beta_s$) shows huge differences to the experimental values, as seen in Fig. 5.30. This suggests a non-Gaussian behavior, which can be described with a relationship of $s = 1.3/\beta_s$. Here it is to mention that by forcing the stretching parameter to be 0.5, i.e., $\beta_s = 0.5$, the data for both temperatures cannot be described with the used fitting function.

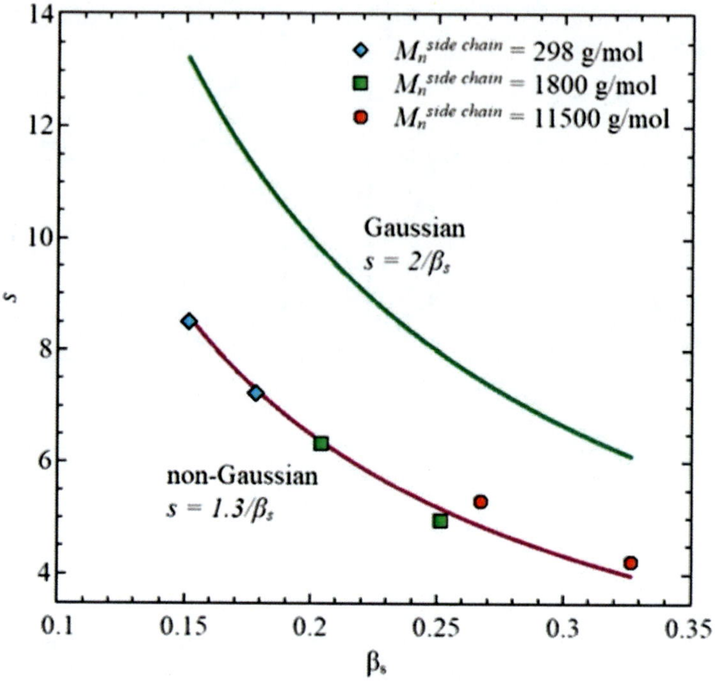

Fig. 5.30 Slope parameter, *s*, vs. stretching parameter of the segmental relaxation, β_s for all three samples. Green solid line represents the relationship for Gaussian behavior, purple solid line represents the data description beyond Gaussian behavior. (*Bichler, K. J., Jakobi, B., García Sakai, V., Klapproth, A., Mole, R. A., Schneider, G. J., Short Time Dynamics of PDMS-g-PDMS Bottlebrush Polymer Melts Investigated by Quasi-Elastic Neutron Scattering. Macromolecules 2020, 10.1021/acs.macromol.0c01846*)

To verify that the segmental relaxation, obtained by quasi-elastic neutron scattering, belongs to the α-relaxation, as determined by dielectric spectroscopy, a comparison of both techniques is very common. The relaxation times obtained by dielectric spectroscopy together with τ_s for every Q-value are represented in Fig. 5.31 for all three samples.

Here, the same behavior as reported by dielectric spectroscopy is seen in the data obtained by QENS. With increasing temperature, the relaxation times decreases. Even if here only two relaxation times per Q-value are available, the tendency, set by dielectric spectroscopy can still be followed.

However, two points are not enough to determine a VFT temperature dependence. In principle it would be possible to extend the VFT behavior of the dielectric spectroscopy, but this would assume that the low temperature VFT behavior also applies to the intermediate temperature range, where no experimental data are available and requires validity over several orders of magnitude, by far outside the accessible time range. But, indeed, the time scale of the QENS relaxation fits with the segmental relaxation times obtained by dielectric spectroscopy. Therefore, the

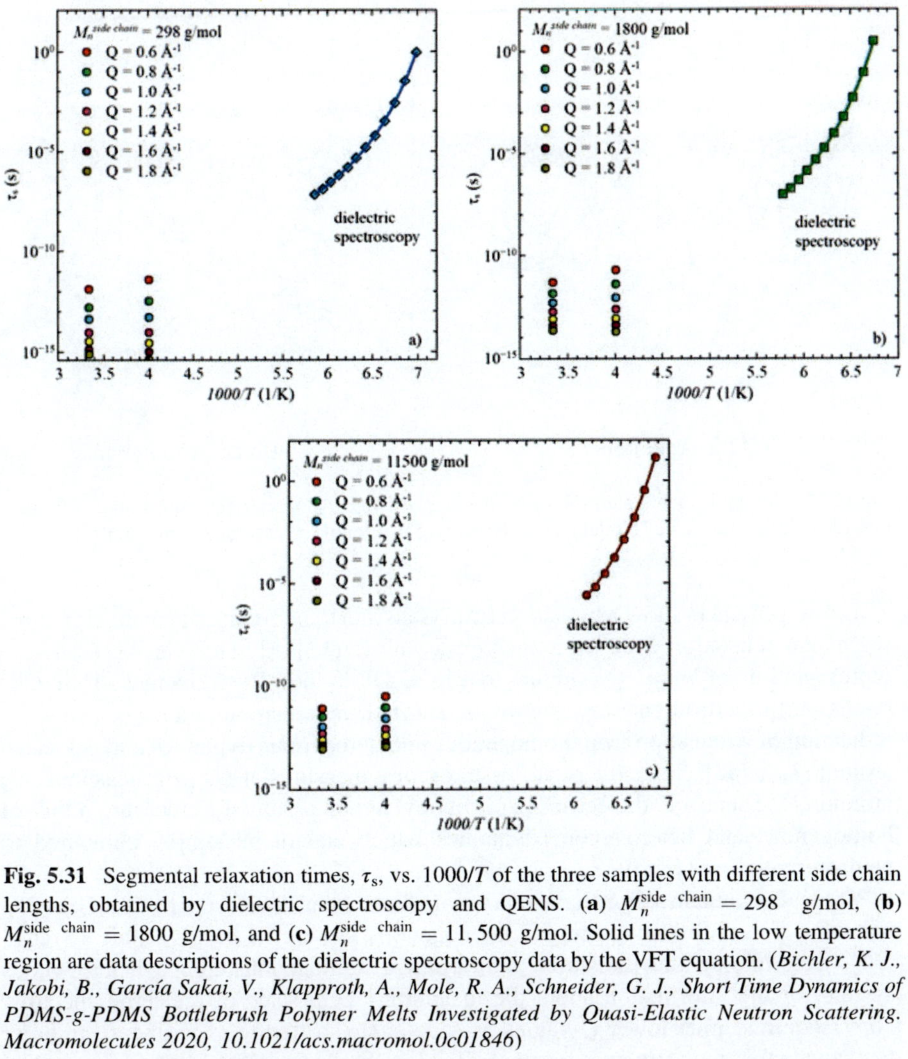

Fig. 5.31 Segmental relaxation times, τ_s, vs. 1000/*T* of the three samples with different side chain lengths, obtained by dielectric spectroscopy and QENS. (**a**) $M_n^{\text{side chain}} = 298$ g/mol, (**b**) $M_n^{\text{side chain}} = 1800$ g/mol, and (**c**) $M_n^{\text{side chain}} = 11,500$ g/mol. Solid lines in the low temperature region are data descriptions of the dielectric spectroscopy data by the VFT equation. (*Bichler, K. J., Jakobi, B., García Sakai, V., Klapproth, A., Mole, R. A., Schneider, G. J., Short Time Dynamics of PDMS-g-PDMS Bottlebrush Polymer Melts Investigated by Quasi-Elastic Neutron Scattering. Macromolecules 2020, 10.1021/acs.macromol.0c01846*)

second relaxation which is determined by QENS is identified as the segmental relaxation of the samples.

Regarding a direct one to one comparison of dielectric spectroscopy with QENS, one big question is the length scale, which is attributed to dielectric spectroscopy. In literature, two different statements about this can be found. The first one states that dielectric spectroscopy measures length scales, associated with the *Q*-value of $Q = 1.0\ \text{Å}^{-1}$ [1]. The other statement is connected to the structural peaks, determined by diffraction experiments. Here, the α-relaxation of dielectric spectroscopy is associated with the Q_{max}-value of the first structure peak [9, 47].

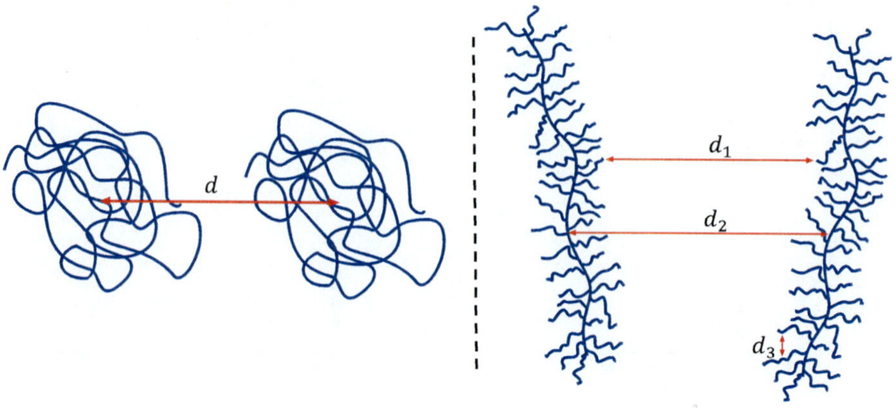

Fig. 5.32 Comparison of intermolecular distances of linear polymer (*left side*) and exemplary intermolecular distances of bottlebrush polymer (*right side*) of the assumed correlations

Linear polymers show a transition from Gaussian to non-Gaussian behavior of the segmental relaxation time by a specific Q-value, Q^* [48]. This can be seen in a representation of s vs. Q, similar to Fig. 5.29 if the slope changes from Q^{-4} (Gaussian) to a different slope (non-Gaussian). This transition point, Q^*, is also an indication of a transition from homogenous to heterogenous dynamics and is located around Q_{max} [48]. Since dielectric spectroscopy measures at Q_{max} it measures also around Q^*. Therefore, dielectric spectroscopy can be assumed to measure a mix of homogenous and heterogenous dynamics which cannot be simply compared to neutron scattering [48, 49].

The values obtained for the α-relaxation show clear deviations from a Gaussian behavior (Fig. 5.30) suggesting a heterogenous system in the available Q-range [50]. Determining the transition point from a homogenous into a heterogenous system for the information about the matching length scale of dielectric spectroscopy, requires either lower Q-values or wide-angle scattering experiments, in order to evaluate the static structure factor, $S(Q)$. Usually, linear PDMS shows one peak in $S(Q)$, at around $Q = 0.8$ Å^{-1}, mainly caused by intermolecular interaction, i.e., interaction between two different polymer chains. The associated peak position is proportional to the distance of two chains by $d = 2\pi/Q_{\mathrm{max}}$ [51].

In case of bottlebrush polymers, it is very difficult to determine an appropriate intermolecular interaction distance which would be connected to the length scale of dielectric spectroscopy. Due to the architecture of the branched polymer, the system should have at least three different intermolecular interactions as illustrated in Fig. 5.32.

Therefore, a simple conclusion about the length scale measured by dielectric spectroscopy cannot be made and the two techniques cannot be easily compared. This would only be possible for smaller Q-values if the system would surpass the Q^* value.

5.2.2.2.4 Mean Square Displacement (MSD)

The mean square displacement (MSD), $\langle r^2(t)\rangle$, is a model independent way to gain information about the dynamics in the systems over the complete temperature range and all relaxations. Since the presented systems are heterogenous within the available Q-window, i.e., the systems have a non-Gaussian behavior, Eq. (2.31) has been used to determine $\langle r^2(t)\rangle$. To recall Eq. (2.31) it is written here (without equation number)

$$S_{\text{inc}}(Q,t) = A\exp\left(-\frac{\langle r^2(t)\rangle}{6}Q^2 + \frac{\langle r^2(t)\rangle^2\alpha_2(t)}{72}Q^4\right)$$

whereby $\alpha_2(t)$ represents the non-Gaussian parameter, which arises from dynamical heterogeneity and describes deviations from a Gaussian behavior [52].

To extract the MSD from the intermediate scattering function, the representation shown in Fig. 5.33 was chosen and described with Eq. (2.31), whereby all-time

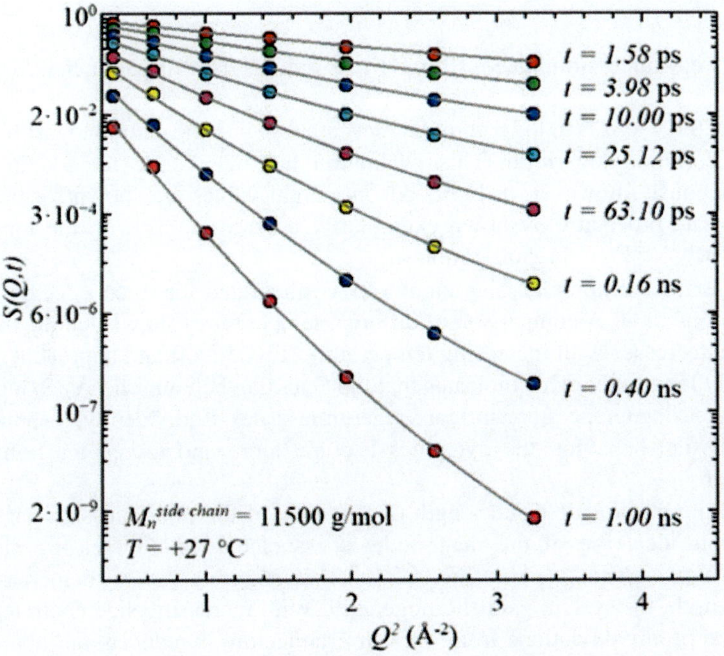

Fig. 5.33 Intermediate Scattering function, $S(Q,t)$, vs. Q^2 for eight selected time values for the PDMS-g-PDMS bottlebrush polymer with the longest side chain length at temperature $T = +27\,°C$. Solid lines are descriptions with Eq. (2.31). (*Bichler, K. J., Jakobi, B., García Sakai, V., Klapproth, A., Mole, R. A., Schneider, G. J., Universality of the Time-Temperature Scaling Observed by Neutron Spectroscopy on Bottlebrush Polymers. Nano Letters 2021, 10.1021/acs.nanolett.1c01379*)

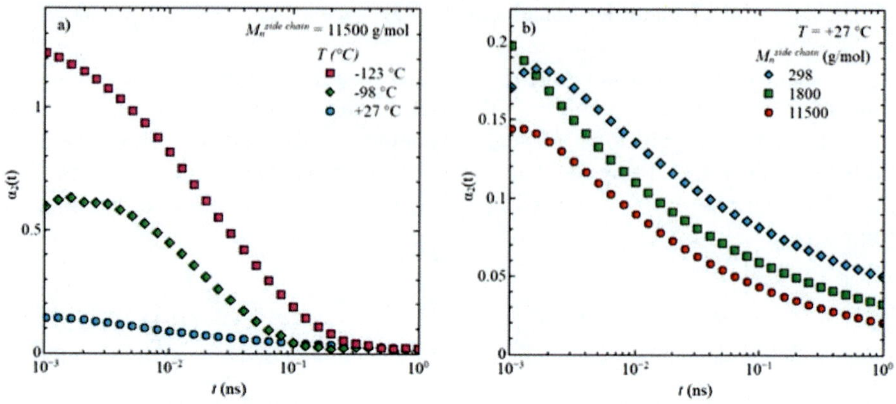

Fig. 5.34 (**a**) Non-Gaussian parameter, α_2, vs. time, t, for the PDMS-g-PDMS bottlebrush polymer with the longest side chain length at three different temperatures, $T = -123, -98$ °C and $T = +27$ °C. (**b**) Non-Gaussian parameter, α_2, vs. time, t, for all three samples at one temperature, $T = +27$ °C. (*Figure (**a**) taken from Bichler, K. J., Jakobi, B., García Sakai, V., Klapproth, A., Mole, R. A., Schneider, G. J., Universality of the Time-Temperature Scaling Observed by Neutron Spectroscopy on Bottlebrush Polymers. Nano Letters 2021, 10.1021/acs.nanolett.1c01379*)

values were fitted simultaneously with two independent fit parameters, $\langle r^2(t)\rangle$ and $\alpha_2(t)$.

Here, the *y*-axis is in logarithmic representation with the natural logarithm instead of the common logarithm. This evaluation has been done for 31 time values simultaneously, however, in Fig. 5.33 only eight values are shown for the sake of clarity. The procedure is shown exemplarily on the sample with the longest side chains, $M_n^{\text{side chain}} = 11,500$ g/mol.

The extracted non-Gaussian parameter is illustrated for three selected temperatures in Fig. 5.34. A comparison of different temperatures shows that the magnitude of $\alpha_2(t)$ decreases with increasing temperature (Fig. 5.34a) and approaches zero for infinitely long times. This indicates that the Gaussian behavior becomes more likely with increasing temperature and for longer time scales. Equivalently, by approaching the Gaussian behavior, the dynamics become more and more of a homogenous character.

Looking at the side chain length dependence of the non-Gaussian parameter, a continuous decrease of the magnitude is associated with increasing side chain molecular weight as seen in Fig. 5.34b. This suggests that with increasing side chain length, the system gains homogeneity. With increasing side chain length, the influence of any deviations from a linear architecture is reduced and approaches a behavior similar to linear polymer chains. This could be seen as the reason for the lower magnitude of the non-Gaussian parameter for longer side chains.

The second fit parameter obtained from Eq. (2.31) is $\langle r^2(t)\rangle$. Hereby, the mean square displacement represents all length scales associated with the available Q-range. For all three polymers, $\langle r^2(t)\rangle$ as a function of time, t, is shown in Fig. 5.35.

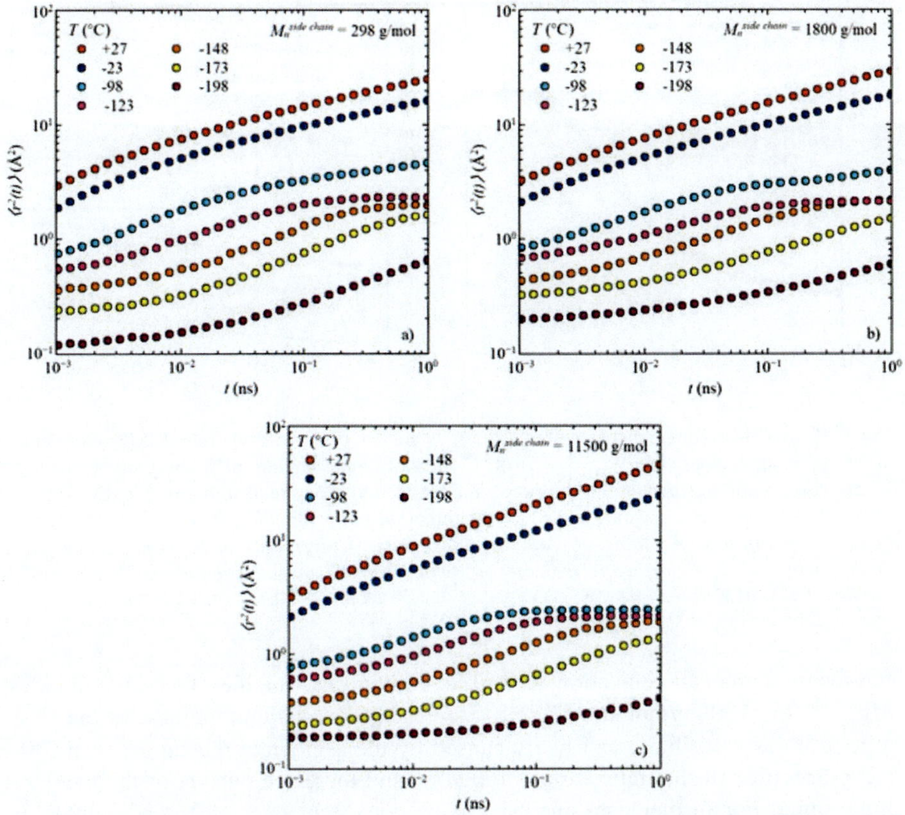

Fig. 5.35 Mean square displacement, $\langle r^2(t)\rangle$, vs. time, t, of PDMS-g-PDMS bottlebrush polymers for different temperatures. (**a**) $M_n^{\text{side chain}} = 298$ g/mol, (**b**) $M_n^{\text{side chain}} = 1800$ g/mol and (**c**) $M_n^{\text{side chain}} = 11,500$ g/mol. For further data treatment, $T = -98$ °C in case (**a**) and (**b**) is excluded. (*Bichler, K. J., Jakobi, B., García Sakai, V., Klapproth, A., Mole, R. A., Schneider, G. J., Universality of the Time-Temperature Scaling Observed by Neutron Spectroscopy on Bottlebrush Polymers. Nano Letters 2021, 10.1021/acs.nanolett.1c01379*)

Two different regions can be distinguished. At the high temperatures the MSD increases with increasing temperature, whereby at low temperatures, the MSD approaches a plateau in the long-time limit. This is consistent with the two different relaxation processes, observed in the intermediate scattering function.

The high temperatures belong to the segmental dynamics, whereas the low temperatures represent the methyl group reorientation. At long times, the restricted dynamics of the threefold jump of the hydrogen atoms causes an apparent plateau.

The general behavior in the MSD is similar for all three samples. However, a more detailed inspection reveals differences based on the highest temperature of the methyl group reorientation, $T = -98$ °C. In case of the bottlebrush polymers, having the longest side chains, $M_n^{\text{side chain}} = 11,500$ g/mol, no contribution of the

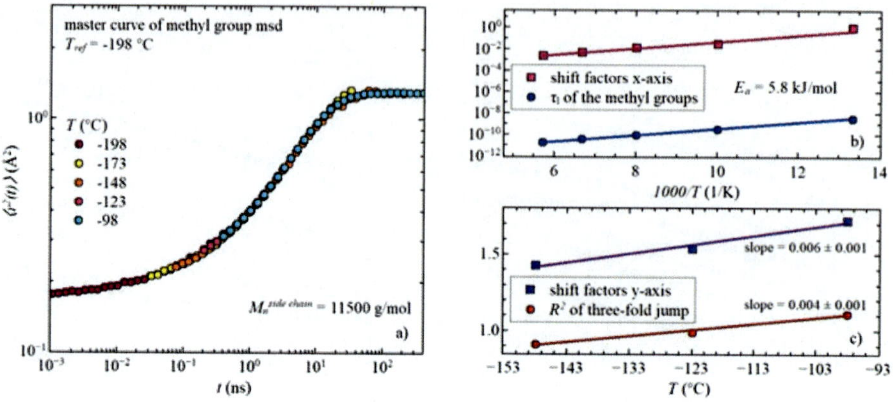

Fig. 5.36 (**a**) Mean square displacement, $\langle r^2(t)\rangle$, vs. time, t, of the methyl group MSD master curve at the reference temperature, $T_{\text{ref}} = -198$ °C for the sample having the longest side chains. (**b**) Comparison of the shift factors for the x-axis and the methyl group relaxation time. (**c**) Comparison of the shift factors for the y-axis and R^2 of the threefold jump. (*Bichler, K. J., Jakobi, B., García Sakai, V., Klapproth, A., Mole, R. A., Schneider, G. J., Universality of the Time-Temperature Scaling Observed by Neutron Spectroscopy on Bottlebrush Polymers. Nano Letters 2021, 10.1021/acs.nanolett.1c01379*)

following segmental relaxation is visible. In contrast, for the other two samples, $M_n^{\text{side chain}} = 1800$ g/mol and $M_n^{\text{side chain}} = 298$ g/mol, a slight increase of the MSD at long times is visible, most likely caused by the segmental relaxation even if the relaxation time of that relaxation is too slow and therefore outside of the available time range. For further data analysis, to be consistent over all three samples the temperature $T = -98$ °C is excluded for the samples with $M_n^{\text{side chain}} = 1800$ g/mol and $M_n^{\text{side chain}} = 298$ g/mol.

For the following data analysis, the Time–Temperature Superposition (TTS) principle was used to create a master curve, which significantly extends the available time range and gives an overview over all dynamical regions in the samples. In order to eliminate or minimize any dynamical influences coming from fast vibrations, the atomistic mean square displacement, $\langle u^2\rangle$, as seen in Fig. 5.27 has been subtracted. The adjusted methyl group MSD is shifted (x- and y- direction) to the reference temperature, $T = -198$ °C, resulting in the methyl group master curve as seen in Fig. 5.36a. Here the time window has been extended from originally three to over five orders of magnitude in time.

The used shift parameters for the x-axis, shown in Fig. 5.36b, have the same slope as the relaxation times of the methyl group dynamics and can therefore be described by the same activation energy E_a. Furthermore, the shift parameters for the y-direction need to show similar behavior as the square radius, R^2, of the methyl groups as seen in Fig. 5.36c. The offset between the shift parameter and the related quantity stems only from the different nature of the values. Only the matching slope

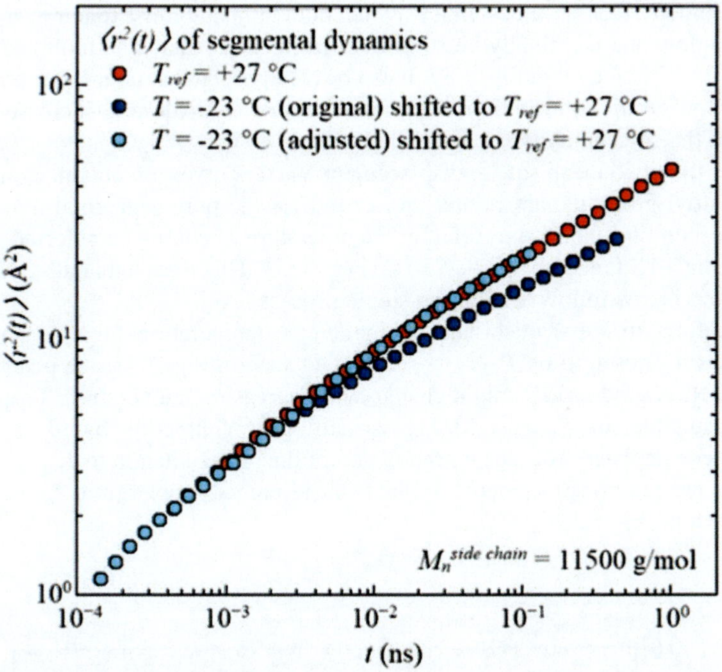

Fig. 5.37 Mean square displacement, $\langle r^2(t) \rangle$, vs. time, t, for temperatures $T = +27$ °C and $T = -23$ °C for the PDMS-g-PDMS bottlebrush with the longest side chains. (*Bichler, K. J., Jakobi, B., García Sakai, V., Klapproth, A., Mole, R. A., Schneider, G. J., Universality of the Time-Temperature Scaling Observed by Neutron Spectroscopy on Bottlebrush Polymers. Nano Letters 2021, 10.1021/acs.nanolett.1c01379*)

is important and confirms the validity of the underlying time–temperature superposition principle.

In order to create a master curve of the segmental dynamics, $T = -23$ °C was shifted to $T = +27$ °C. At that temperature, only the segmental relaxation is observable. However, only shifting $T = -23$ °C without any modification does not lead to match with $T = +27$ °C (Fig. 5.37). In principle, according to the TTS, if both temperatures only contain the segmental dynamics, both should coincide. For $T = -23$ °C an adjustment for the methyl group mean square displacement was implemented, according to:

$$\mathrm{msd}\left(T = -23^{\circ}\mathrm{C},\ \text{segment only}\right) = \mathrm{msd}\left(T = -23^{\circ}\mathrm{C}\right) - \mathrm{msd}\left(T = -98^{\circ}\mathrm{C}\right) \tag{5.15}$$

For that, the highest temperature of the methyl groups MSD, in case of the sample with the longest side chains $T = -98$ °C (Fig. 5.35c), was shifted to $T = -23$ °C with the shift parameter determined based on the Arrhenius behavior. The plateau was extended to longer times, in a way that both temperatures expand over the same time window and can finally be subtracted from each other. The modified MSD of $T = -23$ °C is shown in Fig. 5.37. It can be clearly seen that after the subtraction of the methyl group contribution, both temperatures coincide and increase the time range of the segmental dynamics MSD.

After that, two mean square displacement master curves are obtained, one for the pure methyl group dynamics and the second for the pure segmental dynamics. In order to combine both, one reference temperature needs to be selected. As seen earlier, the TTS failed at $T = -23$ °C (Fig. 5.37). This means that all temperatures above and below follow TTS at the single process level.

Therefore, to solve this issue, the reference temperature for connecting both processes is chosen to be $T_{ref} = -23$ °C. As seen in Fig. 5.35, the plateau height of the methyl group MSD might change with increasing temperature. To adjust this height, suitable for T_{ref}, the MSD was shifted in y-direction based on the shift parameter calculated from the extrapolation of the plateau height to $T_{ref} = -23$ °C. Both mean square displacements shifted to the reference temperature, $T_{ref} = -23$ °C, are shown in Fig. 5.38.

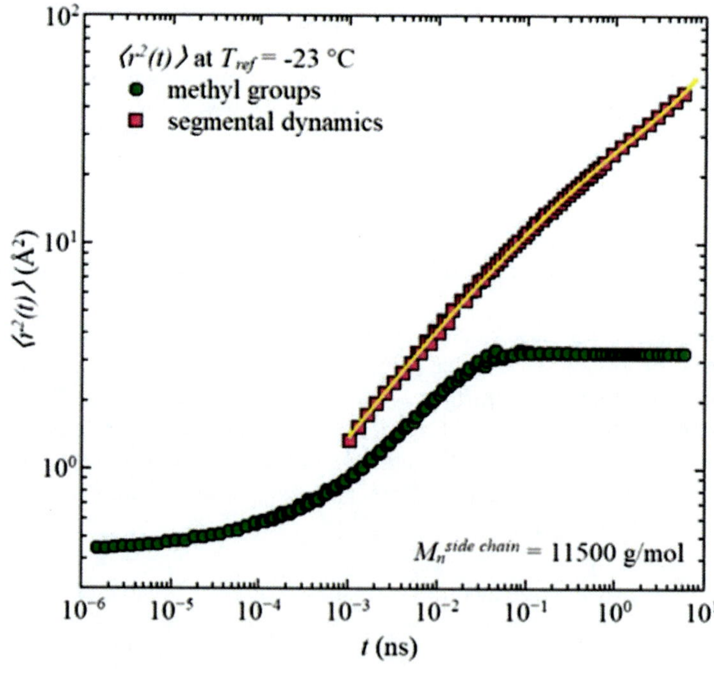

Fig. 5.38 Mean square displacement, $\langle r^2(t) \rangle$, vs. time, t, of the two separated processes, methyl group and segmental dynamics for the PDMS-g-PDMS bottlebrush with the longest side chains at reference temperature $T_{ref} = -23$ °C. Solid yellow line represents an interpolation for the master curve of the segmental MSD

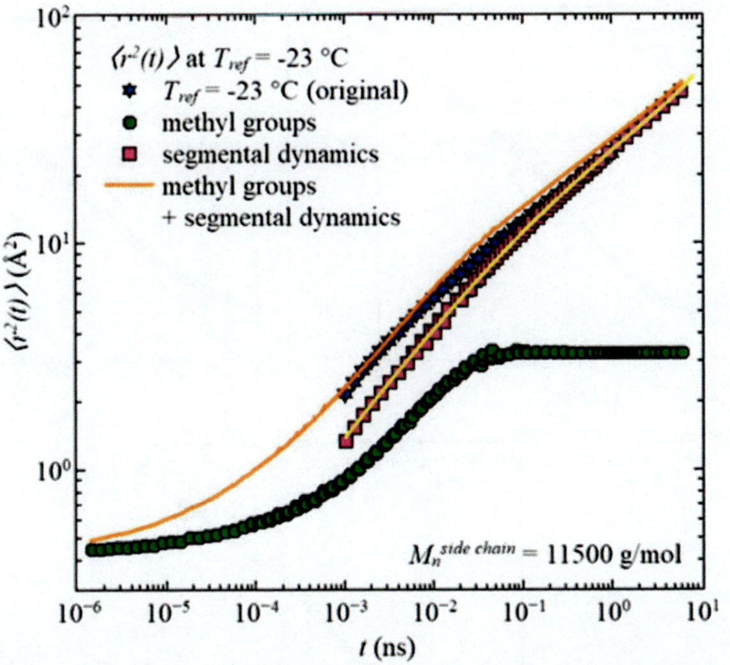

Fig. 5.39 Mean square displacement, $\langle r^2(t)\rangle$, vs. time, t, for the PDMS-g-PDMS bottlebrush with the longest side chains at reference temperature $T_{\text{ref}} = -23$ °C. Solid yellow line represents an interpolation for the master curve of the segmental MSD. Solid orange line represents the combine MSD containing methyl group and segmental contribution. (*Bichler, K. J., Jakobi, B., García Sakai, V., Klapproth, A., Mole, R. A., Schneider, G. J., Universality of the Time-Temperature Scaling Observed by Neutron Spectroscopy on Bottlebrush Polymers. Nano Letters 2021, 10.1021/ acs.nanolett.1c01379*)

Additionally, the MSD of the segmental dynamics was interpolated by a sum of power laws (yellow line in Fig. 5.38).

$$\langle r^2(t)\rangle = ax^b + a_1x^{b_1} + a_2x^{b_2} \tag{5.16}$$

Here it is to mention that the most important aspect of the interpolation was not the physical correctness. Instead, the aim was to find a parametrization of the segmental mean square displacement for further data treatment, allowing to create the master curve of the segmental mean square displacement in the same time range with exactly the same points as obtained for the methyl group MSD master curve.

Summation of the segmental and the methyl group MSD over the entire time range, which expands over seven decades, results in a combined MSD, containing both, segmental and methyl groups contributions, as shown as the solid orange line in Fig. 5.39.

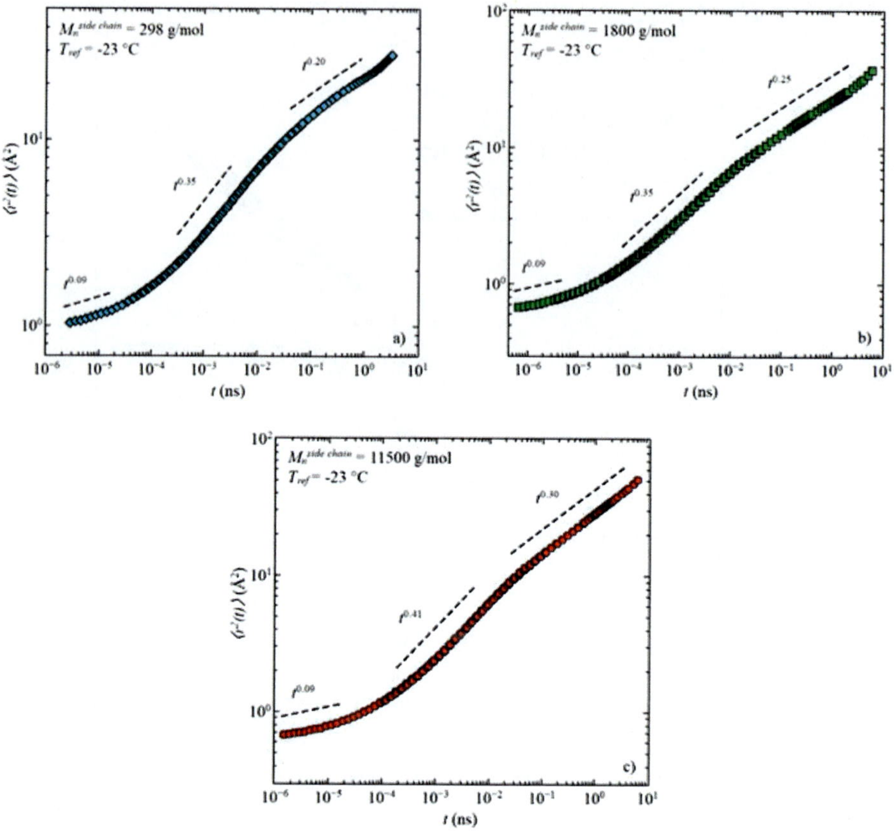

Fig. 5.40 Final mean square displacement, $\langle r^2(t)\rangle$, vs. time, t, for all samples. Three samples, **(a)** $M_n^{\text{side chain}} = 298$ g/mol, **(b)** $M_n^{\text{side chain}} = 1800$ g/mol, and **(c)** $M_n^{\text{side chain}} = 11,500$ g/mol. The standard deviations of the power laws are <1%. (*Bichler, K. J., Jakobi, B., García Sakai, V., Klapproth, A., Mole, R. A., Schneider, G. J., Universality of the Time-Temperature Scaling Observed by Neutron Spectroscopy on Bottlebrush Polymers. Nano Letters 2021, 10.1021/acs.nanolett.1c01379*)

Here the Debye–Waller factor is not included yet. After the pointwise summation of the two single processes, the original MSD of the reference temperature, $T_{\text{ref}} = -23$ °C can be, within the experimental error, described in a proper way. This supports the argument that $T = -23$ °C still has contributions from the methyl groups included. Since the fitting results of the intermediate scattering function imply only segmental relaxation at that temperature, it can be assumed that the methyl group contribution cannot be resolved by means of the intermediate scattering function.

Adding the Debye–Waller factor, for $T = -23$ °C to the combined MSD of segmental and methyl group dynamics, results in the final MSD containing all observed contributions from QENS as seen in Fig. 5.40.

In all three samples, three different regions are observed. All samples show the same power law at short times, $\langle r^2(t)\rangle \propto t^{0.09}$, however, the intermediate and long-time scale show different power laws, changing in a systematic order, depending on the side chain length. While in the intermediate region, the power law goes from $\langle r^2(t)\rangle \propto t^{0.35}$ to $\langle r^2(t)\rangle \propto t^{0.41}$, the change on the long-time scale reaches from $\langle r^2(t)\rangle \propto t^{0.20}$ to $\langle r^2(t)\rangle \propto t^{0.30}$ with increasing molecular weight of the side chains. Comparing absolute values of the MSD for the three different samples, it reveals that the sample with the longest side chains has the weakest restriction and maximal movement in the same time compared to the other two samples.

However, all these power laws cannot be attributed to the power laws known for the MSD of linear polymers (Fig. 2.4). Since this is a new kind of data treatment, especially for the final mean square displacement, and additionally, nothing can be found in literature about the mean square displacement of branched, or bottlebrush polymers, measured in melt, it is hard to predict any behavior based on the power laws. Support by molecular dynamic simulation studies would be needed to gain additional insight into the dynamics of bottlebrush polymers. Also, future experiments on the same samples, as introduced here, with the fast field cycling technique should allow to extend the time range, accessible by QENS. This allows to cover times up to the center of mass diffusion and may simplify the assignment of time scales to specific dynamical regimes.

References

1. F. Kremer, A. Schönhals, *Broadband Dielectric Spectroscopy* (Springer, Berlin, Heidelberg, 2012)
2. G.R. Strobl, *The Physics of Polymers: Concepts for Understanding Their Structures and Behavior* (Springer, Berlin, Heidelberg, 2013)
3. R. Díaz-Calleja, Comment on the maximum in the loss permittivity for the Havriliak–Negami equation. Macromolecules **33**(24), 8924–8924 (2000)
4. A. Boersma, J. Turnhout, M. Wübbenhorst, Dielectric characterization of a thermotropic liquid crystalline copolyesteramide: 1. Relaxation peak assignment. Macromolecules **31**, 7453–7460 (1998)
5. C. Gerstl, G.J. Schneider, W. Pyckhout-Hintzen, J. Allgaier, D. Richter, A. Alegría, J. Colmenero, Segmental and normal mode relaxation of poly(alkylene oxide)s studied by dielectric spectroscopy and rheology. Macromolecules **43**(11), 4968–4977 (2010)
6. K. Xing, S. Chatterjee, T. Saito, C. Gainaru, A.P. Sokolov, Impact of hydrogen bonding on dynamics of hydroxyl-terminated polydimethylsiloxane. Macromolecules **49**(8), 3138–3147 (2016)
7. A. Herrmann, B. Kresse, J. Gmeiner, A.F. Privalov, D. Kruk, F. Fujara, E.A. Rössler, Protracted crossover to reptation dynamics: A field cycling 1H NMR study including extremely low frequencies. Macromolecules **45**(3), 1408–1416 (2012)
8. B. Kresse, M. Hofmann, A.F. Privalov, N. Fatkullin, F. Fujara, E.A. Rössler, All polymer diffusion regimes covered by combining field-cycling and field-gradient 1H NMR. Macromolecules **48**(13), 4491–4502 (2015)
9. C. Gerstl, G.J. Schneider, A. Fuxman, M. Zamponi, B. Frick, T. Seydel, M. Koza, A.C. Genix, J. Allgaier, D. Richter, J. Colmenero, A. Arbe, Quasielastic neutron scattering study on the dynamics of poly(alkylene oxide)s. Macromolecules **45**(10), 4394–4405 (2012)

10. J. Colmenero, A. Arbe, Recent progress on polymer dynamics by neutron scattering: From simple polymers to complex materials. J. Polym. Sci. B Polym. Phys. **51**(2), 87–113 (2013)
11. V. Arrighi, F. Ganazzoli, C. Zhang, S. Gagliardi, New interpretation of local dynamics of poly (dimethyl siloxane) observed by quasielastic neutron scattering. Phys. Rev. Lett. **90**(5), 058301 (2003)
12. V. Arrighi, S. Gagliardi, C. Zhang, F. Ganazzoli, J.S. Higgins, R. Ocone, M.T.F. Telling, A unified picture of the local dynamics of poly(dimethylsiloxane) across the melting point. Macromolecules **36**(23), 8738–8748 (2003)
13. V. Arrighi, J.S. Higgins, A.N. Burgess, W.S. Howells, Rotation of methyl side groups in polymers: A Fourier transform approach to quasielastic neutron scattering. 1. Homopolymers. Macromolecules **28**(8), 2745–2753 (1995)
14. R. Zorn, B. Frick, L. Fetters, Quasielastic neutron scattering study of the methyl group dynamics in polyisoprene. J. Chem. Phys. **116**(2), 845–853 (2002)
15. V. Arrighi, J. Higgins, Side group rotations in amorphous polymers. Phys. B Condens. Matter **226**(1-3), 1–9 (1996)
16. C. Gerstl, *Kettenkonformation und Dynamik verschiedener Poly(alkylenoxid)e* (Westfälische Wilhlems-Universität, Münster, 2011)
17. J. Colmenero, A.A. Arbe, A. Alegría, K.L. Ngai, Q-dependence pf the relaxation times of the α-relaxation as observed by quasielastic neutron scattering. J. Non-Cryst. Solids **172–174**, 229–233 (1994)
18. J. Colmenero, Segmental dynamics in polymer melts by relaxation techniques and quasielastic neutron scattering. Phys. Scr. **T49A**, 227–232 (1993)
19. K.U. Kirst, F. Kremer, V.M. Litvinov, Broad-band dielectric spectroscopy on the molecular dynamics of bulk and adsorbed poly (dimethylsiloxane). Macromolecules **26**(5), 975–980 (1993)
20. R. Lund, A. Alegria, L. Goitandia, J. Colmenero, M.A. González, P. Lindner, Dynamical and structural aspects of the cold crystallization of poly (dimethylsiloxane)(PDMS). Macromolecules **41**(4), 1364–1376 (2008)
21. H. Adachi, K. Adachi, Y. Ishida, T. Kotaka, Dielectric relaxation of polydimethylsiloxane. J. Polym. Sci. Polym. Phys. Ed. **17**(5), 851–857 (1979)
22. C. Grigoriadis, A. Nese, K. Matyjaszewski, T. Pakula, H.-J. Butt, G. Floudas, Dynamic homogeneity by architectural design—Bottlebrush polymers. Macromol. Chem. Phys. **213** (13), 1311–1320 (2012)
23. C.C. Ibeh, *Thermoplastic Materials: Properties, Manufacturing Methods, and Applications* (CRC, Boca Raton, FL, 2011)
24. B. Tieke, *Makromolekulare Chemie: Eine Einführung* (Wiley-VCH Verlag, Berlin, 2014)
25. D.W. van Krevelen, K. te Nijenhuis, *Properties of Polymers: Their Correlation with Chemical Structure; their Numerical Estimation and Prediction from Additive Group Contributions* (Elsevier Science, Amsterdam, 2009)
26. J. Hintermeyer, A. Herrmann, R. Kahlau, C. Goiceanu, E.A. Rössler, Molecular weight dependence of glassy dynamics in linear polymers revisited. Macromolecules **41**(23), 9335–9344 (2008)
27. A.A. Mansour, B. Stoll, Method of determination of dielectric relaxation characteristics of plasticized polystyrenes on the basis of calorimetric glass temperature. Colloid Polym. Sci. **271** (9), 834–842 (1993)
28. J.E. Mark, *Polymer Data Handbook* (Oxford University Press, New York, 2009)
29. S. Clarson, K. Dodgson, J. Semlyen, Studies of cyclic and linear poly (dimethylsiloxanes): 19. Glass transition temperatures and crystallization behaviour. Polymer **26**(6), 930–934 (1985)
30. S.J. Dalsin, M.A. Hillmyer, F.S. Bates, Linear rheology of polyolefin-based bottlebrush polymers. Macromolecules **48**(13), 4680–4691 (2015)

31. Y. Tsukahara, S.-I. Namba, J. Iwasa, Y. Nakano, K. Kaeriyama, M. Takahashi, Bulk properties of poly(macromonomer)s of increased backbone and branch lengths. Macromolecules **34**(8), 2624–2629 (2001)
32. C.R. López-Barrón, P. Brant, A.P.R. Eberle, D.J. Crowther, Linear rheology and structure of molecular bottlebrushes with short side chains. J. Rheol. **59**(3), 865–883 (2015)
33. C. Roland, K. Ngai, Normalization of the temperature dependence of segmental relaxation times. Macromolecules **25**(21), 5765–5768 (1992)
34. M. Hofmann, A. Herrmann, A. Abou Elfadl, D. Kruk, M. Wohlfahrt, E.A. Rössler, Glassy, rouse, and entanglement dynamics as revealed by field cycling 1H NMR RELAXOMETRY. Macromolecules **45**(5), 2390–2401 (2012)
35. A. Herrmann, B. Kresse, M. Wohlfahrt, I. Bauer, A.F. Privalov, D. Kruk, N. Fatkullin, F. Fujara, E.A. Rössler, Mean square displacement and reorientational correlation function in entangled polymer melts revealed by field cycling 1H and 2H NMR relaxometry. Macromolecules **45**(16), 6516–6526 (2012)
36. S. Kariyo, A. Brodin, C. Gainaru, A. Herrmann, J. Hintermeyer, H. Schick, V.N. Novikov, E.A. Rössler, From simple liquid to polymer melt. glassy and polymer dynamics studied by fast field cycling NMR relaxometry: Rouse Regime. Macromolecules **41**(14), 5322–5332 (2008)
37. A. Herrmann, S. Kariyo, A. Abou Elfadl, R. Meier, J. Gmeiner, V.N. Novikov, E.A. Rössler, Universal polymer dynamics revealed by field cycling 1H NMR. Macromolecules **42**(14), 5236–5243 (2009)
38. M. Rubinstein, R.H. Colby, *Polymer Physics* (OUP, Oxford, 2003)
39. R. Meier, A. Herrmann, M. Hofmann, B. Schmidtke, B. Kresse, A.F. Privalov, D. Kruk, F. Fujara, E.A. Rössler, Iso-frictional mass dependence of diffusion of polymer melts revealed by 1H NMR relaxometry. Macromolecules **46**(14), 5538–5548 (2013)
40. A. Maus, K. Saalwächter, Crystallization kinetics of poly (dimethylsiloxane) molecular-weight blends—Correlation with local chain order in the melt? Macromol. Chem. Phys. **208**(19-20), 2066–2075 (2007)
41. W.F. Daniel, J. Burdynska, M. Vatankhah-Varnoosfaderani, K. Matyjaszewski, J. Paturej, M. Rubinstein, A.V. Dobrynin, S.S. Sheiko, Solvent-free, supersoft and superelastic bottlebrush melts and networks. Nat. Mater. **15**(2), 183–189 (2016)
42. H.G. Elias, *Makromoleküle: Physikalische Struktur und Eigenschaften* (Wiley, New York, 2009)
43. H.G. Elias, *Macromolecules 1: Volume 1: Structure and Properties* (Springer US, Berlin, 2012)
44. C. Zhang, V. Arrighi, S. Gagliardi, I.J. McEwen, J. Tanchawanich, M.T.F. Telling, J.M. Zanotti, Quasielastic neutron scattering measurements of fast process and methyl group dynamics in glassy poly(vinyl acetate). Chem. Phys. **328**(1), 53–63 (2006)
45. H.H. Grapengeter, B. Alefeld, R. Kosfeld, An investigation of micro-brownian motions in polydimethylsiloxane by complementary incoherent-neutron-scattering and nuclear-magnetic-resonance experiments below room temperature. Colloid Polym. Sci. **265**(3), 226–233 (1987)
46. A. Arbe, J. Colmenero, F. Alvarez, M. Monkenbusch, D. Richter, B. Farago, B. Frick, Non-Gaussian nature of the $\ensuremath{\alpha}$ relaxation of glass-forming polyisoprene. Phys. Rev. Lett. **89**(24), 245701 (2002)
47. M. Tyagi, A. Arbe, F. Alvarez, J. Colmenero, M.A. González, Short-range order and collective dynamics of poly(vinyl acetate): A combined study by neutron scattering and molecular dynamics simulations. J. Chem. Phys. **129**(22), 224903 (2008)
48. J. Colmenero, F. Alvarez, A. Arbe, Self-motion and the α relaxation in a simulated glass-forming polymer: Crossover from Gaussian to non-Gaussian dynamic behavior. Phys. Rev. E **65**(4), 041804 (2002)

49. A. Arbe, J. Colmenero, M. Monkenbusch, D. Richter, Dynamics of glass-forming polymers:"Homogeneous" versus "heterogeneous" scenario. Phys. Rev. Lett. **81**(3), 590 (1998)
50. M. Tyagi, A. Alegría, J. Colmenero, Heterogeneous dynamics of poly(vinyl acetate) far above Tg: A combined study by dielectric spectroscopy and quasielastic neutron scattering. J. Chem. Phys. **122**(24), 244909 (2005)
51. V. Arrighi, J.S. Higgins, A.N. Burgess, G. Floudas, Local dynamics of poly(dimethyl siloxane) in the presence of reinforcing filler particles. Polymer **39**(25), 6369–6376 (1998)
52. T. Kanaya, I. Tsukushi, K. Kaji, Non-Gaussian parameter and heterogeneity of amorphous polymers. Prog. Theor. Phys. Suppl. **126**, 133–140 (1997)

Chapter 6
Summary and Conclusion

Abstract The morphological and dynamical behaviors of PDMS-g-PDMS bottlebrush polymers were investigated by different techniques. Structural changes induced by varying the ratio from side chain to backbone molecular weight or by changing the host environment were followed by small-angle neutron scattering. Since polymer dynamics spread over several orders of magnitude in time, a combination of instruments was used to get information about the dynamical process in the time range from pico to micro seconds. Segmental dynamics over several temperatures were tracked by dielectric spectroscopy and fast field cycling relaxometry. Hereby, the latter method has additional access to the large-scale polymer dynamics expressed by the mean square displacement. For information about the length scale dependent relaxation behavior, quasi-elastic neutron scattering was used covering the short-time relaxation behavior, including methyl group rotation and segmental dynamics.

This dissertation investigates the morphology and dynamics of bottlebrush polymers. This type of polymer has linear side chains covalently bonded on a linear backbone, resembling the so-called bottlebrush structure. Along with that, extraordinary properties, e.g., super softness or low viscosities, can be found. To understand these macroscopic properties, first, the microscopic origins need to be investigated. Therefore, neutron scattering and relaxometry experiments have been performed, which are well suited for the length and time scales of polymers, i.e., morphology and dynamics. All the used samples are based on PDMS, and well customized regarding the molecular weight, the dispersity, and the isotopic labeling, making them ideally suited for neutron scattering experiments.

Structural investigations on four different PDMS based samples systems, i.e., spherical PDMS-g-PDMS bottlebrush in good solvent, in good solvent with added linear chains, in linear melt of varying molecular weight, and PDMS-g-PDMS bottlebrush polymers performing the shape transition in good solvent, have been done by Small-Angle Neutron Scattering (SANS) experiments.

For dynamical studies, a comparison between the relaxation behavior of the spherical bottlebrush polymer to its respective single PDMS side chain has been performed by dielectric spectroscopy. To connect polymer dynamics with

K.J. Bichler, *Morphology and Dynamics of Bottlebrush Polymers*, Springer Theses,
https://doi.org/10.1007/978-3-030-83379-4_6

architectures, the relaxation behavior of the spherical bottlebrush and a linear PDMS with similar molecular weight has been investigated and compared by means of Fast Field Cycling (FFC) relaxometry. For information about the dynamical influences inherent with structural changes, the sample system performing the shape transition has been analyzed by Quasi-Elastic Neutron Scattering (QENS), supplemented by dielectric spectroscopy experiments.

The morphology of PDMS-g-PDMS bottlebrush polymers was investigated under good solvent condition by SANS. A spherical shape has been determined in the low concentration region, whereby with increasing volume fraction a correlation peak evolves. The maximum of this peak is related to the center of mass distance of next neighbored bottlebrushes and allows to specify an experimental overlap concentration. The addition of linear PDMS chains, under zero contrast condition, results in a vanishing correlation peak at high volume fractions (second sample system). This is caused by internal density fluctuations which are generated by the single linear chains, occupying space between the side chains. The third sample system consists of a complete melt sample, spherical PDMS-g-PDMS bottlebrushes in linear PDMS melt, i.e., solvent free system. To follow structural changes caused by the influences of the host matrix, three different matrices have been used, differing in their molecular weight. In the low concentration, the spherical shape is maintained over all linear matrices, while with increasing molecular weight of the linear host matrix, the radius of the bottlebrush decreases caused by the osmotic pressure. However, in the largest molecular weight matrix, at the highest concentration, the formation of clusters sets in. This can be seen as a model system for polymer grafted nanoparticles in relation to polymer nanocomposites for insights about the miscibility in different linear polymer matrices.

The shape of bottlebrush polymers strongly depends on the backbone to side chain size ratio. Based on literature, the so-called sphere to cylinder transition is known for relatively stiff bottlebrush polymers, with poly(norbornene) or poly (styrene) backbone. This shape transition is also found for the flexible PDMS-g-PDMS bottlebrush polymers in good solvent condition, having similar backbone length, but different lengths of side chains. These samples are very interesting for dynamical studies and were further used for QENS and dielectric spectroscopy experiments.

Dynamical studies have been performed on different samples, which prior have been investigated by SANS, in most cases. The comparison of the spherical PDMS-g-PDMS bottlebrush polymer with its respective linear PDMS side chains has been done by dielectric spectroscopy, resulting in the segmental relaxation, i.e., the α-relaxation, which shows two changes. First the relaxation times slowdown in case of the bottlebrush polymer and second the polymer specific glass transition temperature increases. However, independent measurements of the single side chains (linear PDMS), prior to the grafting process show that the detected signal of the PDMS-g-PDMS bottlebrush polymers in dielectric spectroscopy is governed by their side chains.

Since different architectures of PDMS show a reduced segmental relaxation, fast field cycling relaxometry has been used to connect polymer dynamics to different conformations. Hereby, the spherical PDMS-g-PDMS bottlebrush polymer was compared to a linear PDMS with a similar molecular weight as the overall bottlebrush. Based on the extracted mean square displacement from FFC data, the bottlebrush polymer shows Rouse dynamics, followed by diffusion. In contrast, the high molecular weight linear PDMS indicates Rouse dynamics and constraint Rouse dynamics in the available time range. This suggests missing entanglement effects in the bottlebrush polymer even if the molecular weight is high, which is connected with a shifted entanglement molecular weight.

As changing the size ratio from the backbone to the side chains affects the morphology, the impact on the dynamics has been studied by dielectric spectroscopy and quasi-elastic neutron scattering. Herefore, the samples performing the shape transition have been used. In contrast to dielectric spectroscopy, quasi-elastic neutron scattering has not only a high energy resolution but also a spatial resolution connected with the momentum transfer. To capture a broad time range, three instruments have been combined and analyzed in the time domain. This results in time scales of pico to nanoseconds in a size range of $3-10$ Å, capturing methyl group dynamics and segmental dynamics, whereas dielectric spectroscopy has only access to the segmental dynamics. Additionally, the single linear PDMS side chains have been measured by dielectric spectroscopy, supplementing the PDMS-g-PDMS bottlebrush polymers. This experiment resulted in a side chain dominated segmental relaxation of the bottlebrushes, which is slowed done compared to the α-relaxation of the single side chains. This effect depends on the side chain length and saturates at around $M_n^{\text{side chain}} \sim 10,000$ g/mol, as seen on the bottlebrush having side chains of $M_n^{\text{side chain}} = 11,500$ g/mol.

Quasi-elastic neutron scattering experiments on the same samples revealed the methyl group reorientation at the lowest temperatures, and the segmental dynamics at the high temperatures, both with spatial resolution. Hereby, the methyl groups are restricted by their geometry, leading to a description by a threefold jump between the three possible hydrogen positions. Due to the very localized motion, the relaxation times are independent of the momentum transfer and side chain lengths, and are characterized by an Arrhenius temperature behavior. In contrast, the segmental dynamics at the high temperatures show a molecular weight dependence together with segmental relaxation time, showing a well pronounced Q-dependence. This suggests a heterogenous system with a non-Gaussian behavior. Due to the non-Gaussian character, a direct comparison of QENS and dielectric spectroscopy is not possible.

The extraction of the mean square displacement takes the heterogeneity of the system into account by including the non-Gaussian parameter. Unlike the literature, applying the Time–Temperature Superposition (TTS) principle to the two single processes enables to create two mean square displacement master curves. The first one for the pure methyl group dynamics and the second one for the pure segmental dynamics. Both of them are adjusted for fast vibrational motions, which are considered by the Debye-Waller factor. Using TTS for shifting the single master curves to one reference temperature, T_{ref}, and combining both of them, results in the total mean square displacement containing all processes. This procedure extends the original time window from three decades up to seven decades in the final stage. The connected power laws of the mean square displacement show systematic changes, based on the side chain length, but do not follow the power laws, known for linear polymers.

Appendix A: Supplementary Data for the Chapter Morphology

Small-Angle Neutron Scattering of PDMS-g-PDMS Bottlebrush Polymer of $\Phi = 1.0\%$ in Linear PDMS Melt (lin-h-PDMS9k and lin-h-PDMS37k)

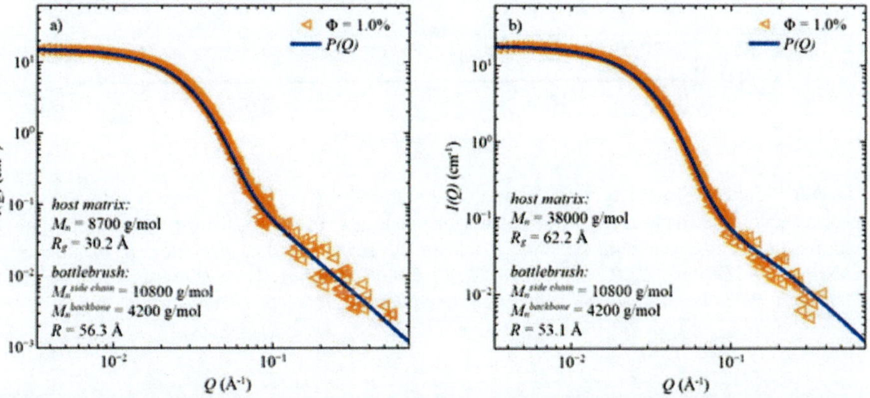

Fig. A.1 Small-angle neutron scattering curve of PDMS-g-PDMS bottlebrush polymer in linear protonated PDMS melt in volume fraction of $\Phi = 1.0\%$. Best fit values are described in Table 4.6. (**a**) Bottlebrushes in lin-h-PDMS9k and (**b**) Bottlebrushes in lin-h-PDMS37k. (*Bichler, K. J., Jakobi, B., Huber, S. O., Gilbert, E. P., Schneider, G. J., Structural Analysis of Ultrasoft PDMS-g-PDMS Shell-Only Particles. Macromolecules 2020,* https://doi.org/10.1021/acs.macromol.9b01598)

K.J. Bichler, *Morphology and Dynamics of Bottlebrush Polymers*, Springer Theses,
https://doi.org/10.1007/978-3-030-83379-4

Simulation Results of Predicted Size Changes in Literature

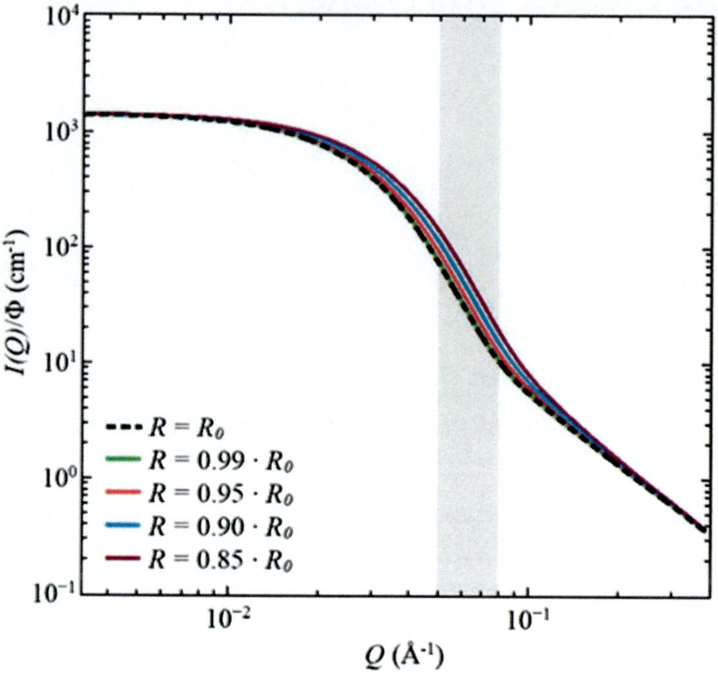

Fig. A.2 Scattering intensity, $I(Q)$, vs. momentum transfer, Q, of the form factor normalized to the respective volume fraction for the simulated size changes. The gray shadowed area highlights the region where changes would become visible in the scattering data. (*Bichler, K. J., Jakobi, B., Huber, S. O., Gilbert, E. P., Schneider, G. J., Structural Analysis of Ultrasoft PDMS-g-PDMS Shell-Only Particles. Macromolecules 2020,* https://doi.org/10.1021/acs.macromol.9b01598)

Appendix B: Supplementary Data for the Chapter Dynamics

Dielectric Spectroscopy of Spherical PDMS-g-PMDS Bottlebrush Polymers

Table B.1 Fit parameter by using the HN-function for describing α-and α_c-relaxation of the PDMS-g-PDMS bottlebrush polymer, including the shape parameter β and γ, as well as the Havriliak–Negami relaxation time, τ_{HN}, and the real relaxation time, τ_s

	Temperature (°C)	β	γ	τ_{HN} (s)	τ_s (s)
α-relaxation	−125.0	0.77	0.51	0.56	0.25575
	−122.5			0.0255	0.0117
	−120.0			0.00211	$9.80 \cdot 10^{-4}$
	−117.5			$2.63 \cdot 10^{-4}$	$1.21 \cdot 10^{-4}$
	−115.0			$4.41 \cdot 10^{-5}$	$2.03 \cdot 10^{-5}$
	−112.5			$8.82 \cdot 10^{-6}$	$4.06 \cdot 10^{-6}$
	−110.0			$2.46 \cdot 10^{-6}$	$1.13 \cdot 10^{-6}$
α_c-relaxation	−107.5	0.47	0.48	0.00011	$2.50 \cdot 10^{-5}$
	−105.0	0.62	0.41	$1.82 \cdot 10^{-5}$	$4.84 \cdot 10^{-6}$
	−102.5	0.61	0.40	$9.74 \cdot 10^{-6}$	$2.44 \cdot 10^{-6}$
	−100.0	0.65	0.36	$3.99 \cdot 10^{-6}$	$9.62 \cdot 10^{-7}$
	-97.5	0.62	0.38	$2.19 \cdot 10^{-6}$	$5.25 \cdot 10^{-7}$
	−95.0	0.58	0.38	$1.34 \cdot 10^{-6}$	$2.79 \cdot 10^{-7}$
	−92.5	0.58	0.37	$9.31 \cdot 10^{-7}$	$1.83 \cdot 10^{-7}$
	−90.0	0.57	0.26	$1.05 \cdot 10^{-6}$	$1.12 \cdot 10^{-7}$
	−87.5	0.56	0.23	$7.90 \cdot 10^{-7}$	$6.72 \cdot 10^{-8}$

All fit values have a standard deviation of around 1% and are omitted for the sake of clarity

K.J. Bichler, *Morphology and Dynamics of Bottlebrush Polymers*, Springer Theses,
https://doi.org/10.1007/978-3-030-83379-4

Single Linear PDMS Side Chain for the Spherical PDMS-g-PMDS Bottlebrush

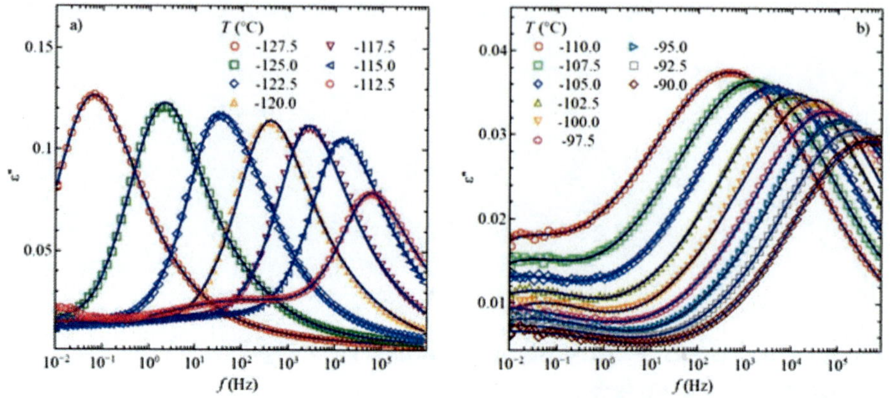

Fig. B.1 Imaginary part of the permittivity, ϵ'', vs. frequency, f, for the single linear PDMS side chains of the spherical PDMS-g-PDMS bottlebrush polymers. (**a**) The segmental relaxation, α-relaxation. (**b**) The segmental relaxation influenced by the cold crystallization, α_c-relaxation. The solid lines are the best description using the Havriliak–Negami function

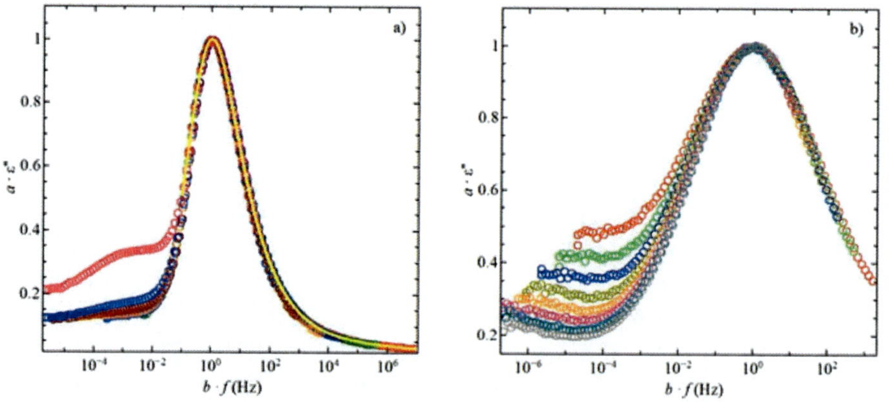

Fig. B.2 Normalized imaginary part of the permittivity, $a \cdot \epsilon''$, vs. normalized frequency, $b \cdot f$, for the single linear PDMS side chains of the spherical PDMS-g-PDMS bottlebrush polymers. (**a**) The segmental relaxation, α-relaxation. (**b**) The segmental relaxation influenced by the cold crystallization, α_c-relaxation

Table B.2 Fit parameter by using the HN-function for describing α-and α_c-relaxation of the single side chains (linear PDMS), including the shape parameter β and γ, as well as the Havriliak–Negami relaxation time, τ_{HN}, and the real relaxation time, τ_s

	Temperature (°C)	β	γ	τ_{HN} (s)	τ_s (s)
α-relaxation	−127.5	0.78	0.56	2.89	1.52
	−125			0.0915	0.0480
	−122.5			0.00538	0.00028
	−120.0			$4.81 \cdot 10^{-4}$	$2.56 \cdot 10^{-5}$
	−117.5			$7.12 \cdot 10^{-5}$	$3.74 \cdot 10^{-6}$
	−115.0			$1.28 \cdot 10^{-5}$	$6.72 \cdot 10^{-7}$
	−112.5			$3.21 \cdot 10^{-6}$	$1.69 \cdot 10^{-7}$
α_c-relaxation	−110.0	0.42	0.57	$2.03 \cdot 10^{-4}$	$5.70 \cdot 10^{-5}$
	−107.5	0.44	0.54	$8.84 \cdot 10^{-5}$	$2.32 \cdot 10^{-5}$
	−105.0	0.46	0.49	$4.84 \cdot 10^{-5}$	$1.11 \cdot 10^{-5}$
	−102.5	0.47	0.50	$1.68 \cdot 10^{-5}$	$4.09 \cdot 10^{-6}$
	−100.0	0.49	0.52	$5.59 \cdot 10^{-6}$	$1.59 \cdot 10^{-6}$
	−97.5	0.48	0.42	$5.99 \cdot 10^{-6}$	$1.04 \cdot 10^{-6}$
	−95.0	0.49	0.44	$2.69 \cdot 10^{-6}$	$5.47 \cdot 10^{-7}$
	−92.5	0.49	0.38	$2.06 \cdot 10^{-6}$	$3.04 \cdot 10^{-7}$
	−90.0	0.47	0.32	$1.69 \cdot 10^{-6}$	$1.69 \cdot 10^{-7}$

All fit values have a standard deviation of around 1% and are omitted for the sake of clarity

Table B.3 Fit parameter of the VFT equation of the single side chains (linear PDSM), describing the temperature dependence of the relaxation time of the α-and α_c-relaxations of the single PDMS side chains including the limiting relaxation time at high temperatures, τ_∞, the fit parameter, A, and the ideal glass transition temperature, T_0

	τ_∞ (s)	A (K)	T_0 (K)
α-relaxation	$(5 \pm 3) \cdot 10^{-15}$	712 ± 27	124.3 ± 0.5
α_c-relaxation	$(7.4 \pm 0.2) \cdot 10^{-12}$	546 ± 171	129 ± 6

Dielectric Spectroscopy of the PDMS-g-PDMS Bottlebrushes (QENS Experiments)

PDMS-g-PDMS Bottlebrush with Short Side Chains, $M_n^{\text{side chain}} = 298$ g/mol

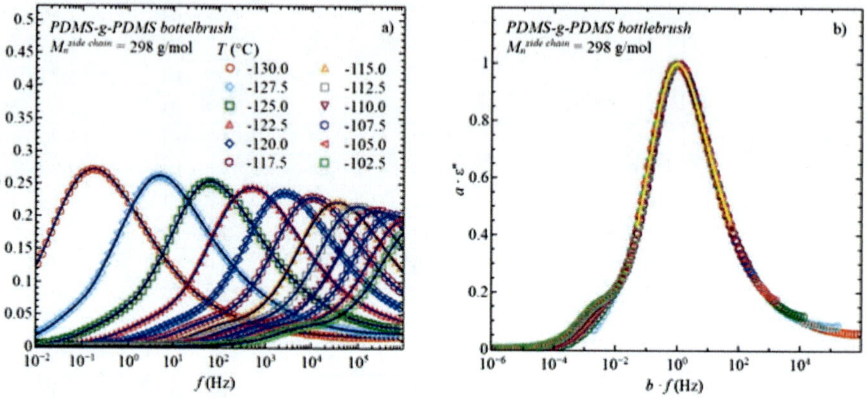

Fig. B.3 (**a**) Imaginary part of the permittivity, ϵ'', vs. frequency, f, of the PDMS-g-PDMS bottlebrush polymer with $M_n^{\text{side chain}} = 298$ for the α-relaxation. Solid lines are the best description with the Havriliak–Negami function. (**b**) Normalized imaginary part of the permittivity, $a \cdot \epsilon''$, vs. normalized frequency, $b \cdot f$, of the PDMS-g-PDMS bottlebrush polymer with $M_n^{\text{side chain}} = 298$ for the α-relaxation. Solid line is the best description with the Havriliak–Negami function. (*Jakobi, B., Bichler, K. J., Sokolova, A., Schneider, G. J., Dynamics of PDMS-g-PDMS Bottlebrush Polymers by Broadband Dielectric Spectroscopy. Macromolecules 2020,* https://doi.org/10.1021/acs.macromol.0c01277)

Table B.4 Fit parameter by using the HN-function for describing α-relaxation of the PDMS-g-PDMS bottlebrush polymer with side chains, $M_n^{\text{side chain}} = 298$ g/mol, including the shape parameter β and γ, as well as the Havriliak–Negami relaxation time, τ_{HN}, and the real relaxation time, τ_s

	Temperature (°C)	β	γ	τ_{HN} (s)	τ_s (s)
α-relaxation	−130.0	0.67	0.48	2.50	0.92
	−127.5			0.09	0.033
	−125.0			$7.14 \cdot 10^{-3}$	$2.63 \cdot 10^{-3}$
	−122.5			$9.00 \cdot 10^{-4}$	$3.32 \cdot 10^{-4}$
	−120.0			$1.69 \cdot 10^{-4}$	$6.22 \cdot 10^{-5}$
	−117.5			$4.28 \cdot 10^{-5}$	$1.58 \cdot 10^{-5}$
	−115.0			$1.30 \cdot 10^{-6}$	$4.79 \cdot 10^{-6}$
	−112.5			$4.50 \cdot 10^{-6}$	$1.66 \cdot 10^{-6}$
	−110.0			$1.79 \cdot 10^{-6}$	$6.60 \cdot 10^{-7}$
	−107.5			$8.16 \cdot 10^{-7}$	$3.01 \cdot 10^{-7}$
	−105.0			$3.57 \cdot 10^{-7}$	$1.32 \cdot 10^{-7}$

(continued)

Table B.4 (continued)

	Temperature (°C)	β	γ	τ_{HN} (s)	τ_s (s)
	−102.5			$1.96 \cdot 10^{-7}$	$7.22 \cdot 10^{-8}$

All fit values have a standard deviation of around 1% and are omitted for the sake of clarity. (*Jakobi, B., Bichler, K. J., Sokolova, A., Schneider, G. J., Dynamics of PDMS-g-PDMS Bottlebrush Polymers by Broadband Dielectric Spectroscopy. Macromolecules 2020,* https://doi.org/10.1021/acs.macromol.0c01277)

Short Single PDMS Side Chains, $M_n^{\text{side chain}} = 298$ g/mol

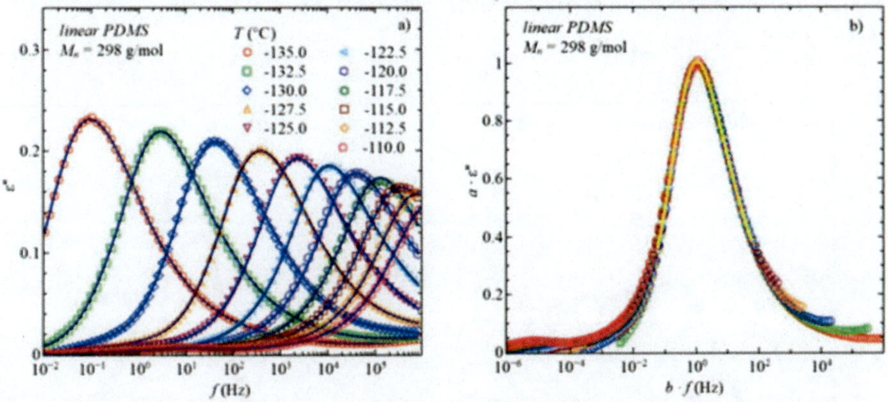

Fig. B.4 (**a**) Imaginary part of the permittivity, ϵ'', vs. frequency, f, of the single linear PDMS side chains with $M_n = 298$ g/mol for the α-relaxation. Solid lines are the best description with the Havriliak–Negami function. (**b**) Normalized imaginary part of the permittivity, $a \cdot \epsilon''$, vs. normalized frequency, $b \cdot f$, of the single linear PDMS side chains with $M_n = 298$ g/mol for the α-relaxation. Solid line is the best description with the Havriliak–Negami function. (*Jakobi, B., Bichler, K. J., Sokolova, A., Schneider, G. J., Dynamics of PDMS-g-PDMS Bottlebrush Polymers by Broadband Dielectric Spectroscopy. Macromolecules 2020,* https://doi.org/10.1021/acs.macromol.0c01277)

Table B.5 Fit parameter by using the HN-function for describing α-relaxation of the linear PDMS side chains, $M_n^{\text{side chain}} = 298$ g/mol, including the shape parameter β and γ, as well as the Havriliak–Negami relaxation time, τ_{HN}, and the real relaxation time, τ_s

	Temperature (°C)	β	γ	τ_{HN} (s)	τ_s (s)
α-relaxation	−135.0	0.74	0.47	4.16	1.68
	−132.5			0.14	0.06
	−130.0			$9.42 \cdot 10^{-3}$	$3.81 \cdot 10^{-3}$
	−127.5			$1.07 \cdot 10^{-3}$	$4.34 \cdot 10^{-4}$
	−125.0			$1.74 \cdot 10^{-4}$	$7.03 \cdot 10^{-5}$
	−122.5			$3.75 \cdot 10^{-5}$	$1.52 \cdot 10^{-5}$
	−120.0			$9.85 \cdot 10^{-6}$	$3.98 \cdot 10^{-6}$
	−117.5			$3.24 \cdot 10^{-6}$	$1.31 \cdot 10^{-6}$

(continued)

Table B.5 (continued)

	Temperature (°C)	β	γ	τ_{HN} (s)	τ_s (s)
	−115.0			$1.19 \cdot 10^{-6}$	$4.80 \cdot 10^{-7}$
	−112.5			$4.19 \cdot 10^{-7}$	$1.99 \cdot 10^{-7}$
	−110.0			$2.35 \cdot 10^{-7}$	$9.51 \cdot 10^{-8}$

All fit values have a standard deviation of around 1% and are omitted for the sake of clarity. (*Jakobi, B., Bichler, K. J., Sokolova, A., Schneider, G. J., Dynamics of PDMS-g-PDMS Bottlebrush Polymers by Broadband Dielectric Spectroscopy. Macromolecules 2020,* https://doi.org/10.1021/acs.macromol.0c01277)

PDMS-g-PDMS Bottlebrush with Middle Side Chains, $M_n^{\text{side chain}} = 1800$ g/mol

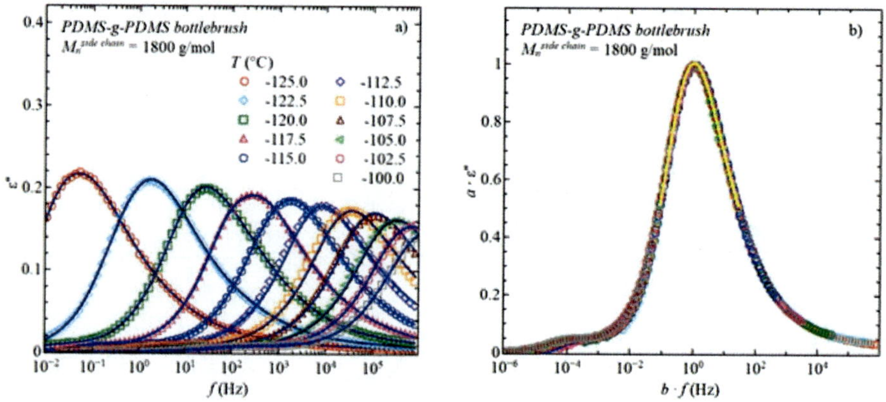

Fig. B.5 (**a**) Imaginary part of the permittivity, ϵ'', vs. frequency, f, of the PDMS-g-PDMS bottlebrush polymer with $M_n^{\text{side chain}} = 1800$ g/mol for the α-relaxation. Solid lines are the best description with the Havriliak–Negami function. (**b**) Normalized imaginary part of the permittivity, $a \cdot \epsilon''$, vs. normalized frequency, $b \cdot f$, of the PDMS-g-PDMS bottlebrush polymer with $M_n^{\text{side chain}} = 1800$ g/mol for the α-relaxation. Solid line is the best description with the Havriliak–Negami function. (*Jakobi, B., Bichler, K. J., Sokolova, A., Schneider, G. J., Dynamics of PDMS-g-PDMS Bottlebrush Polymers by Broadband Dielectric Spectroscopy. Macromolecules 2020,* https://doi.org/10.1021/acs.macromol.0c01277)

Table B.6 Fit parameter by using the HN-function for describing α-relaxation of the PDMS-g-PDMS bottlebrush polymer with side chains, $M_n^{\text{side chain}} = 298$ g/mol, including the shape parameter β and γ, as well as the Havriliak–Negami relaxation time, τ_{HN}, and the real relaxation time, τ_s

	Temperature (°C)	β	γ	τ_{HN} (s)	τ_s (s)
α-relaxation	−125.0	0.69	0.40	8.59	3.30
	−122.5			0.24	0.09
	−120.0			0.01	$5.70 \cdot 10^{-3}$
	−117.5			$1.63 \cdot 10^{-3}$	$6.24 \cdot 10^{-4}$

(continued)

Table B.6 (continued)

	Temperature (°C)	β	γ	τ_{HN} (s)	τ_s (s)
	−115.0			$2.56 \cdot 10^{-4}$	$9.83 \cdot 10^{-5}$
	−112.5			$5.17 \cdot 10^{-5}$	$1.99 \cdot 10^{-5}$
	−110.0			$1.36 \cdot 10^{-5}$	$5.24 \cdot 10^{-6}$
	−107.5			$3.92 \cdot 10^{-6}$	$1.51 \cdot 10^{-6}$
	−105.0			$1.36 \cdot 10^{-6}$	$5.24 \cdot 10^{-7}$
	−102.5			$5.43 \cdot 10^{-7}$	$2.09 \cdot 10^{-7}$
	−100.0			$2.72 \cdot 10^{-7}$	$1.05 \cdot 10^{-7}$

All fit values have a standard deviation of around 1% and are omitted for the sake of clarity. (*Jakobi, B., Bichler, K. J., Sokolova, A., Schneider, G. J., Dynamics of PDMS-g-PDMS Bottlebrush Polymers by Broadband Dielectric Spectroscopy. Macromolecules 2020,* https://doi.org/10.1021/acs.macromol.0c01277)

Middle Single PDMS Side Chains, $M_n^{\text{side chain}} = 1800$ ***g/mol***

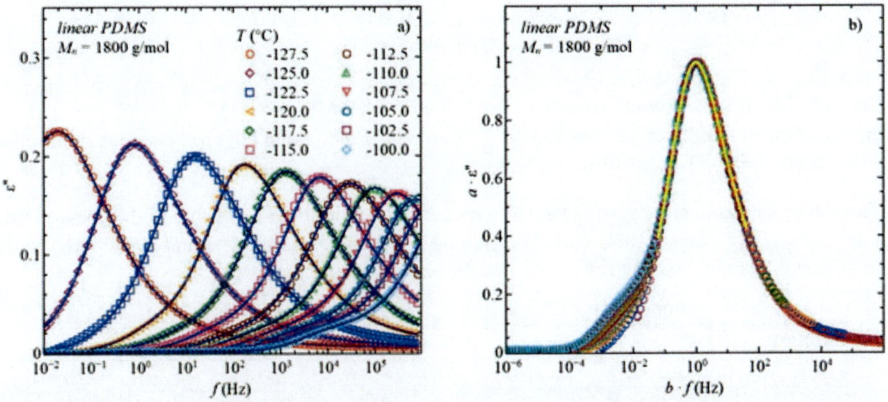

Fig. B.6 (**a**) Imaginary part of the permittivity, ϵ'', vs. frequency, f, of the single linear PDMS side chains with $M_n = 1800$ g/mol for the α-relaxation. Solid lines are the best description with the Havriliak–Negami function. (**b**) Normalized imaginary part of the permittivity, $a \cdot \epsilon''$, vs. normalized frequency, $b \cdot f$, of the single linear PDMS side chains with $M_n = 1800$ g/mol for the α-relaxation. Solid line is the best description with the Havriliak–Negami function. (*Jakobi, B., Bichler, K. J., Sokolova, A., Schneider, G. J., Dynamics of PDMS-g-PDMS Bottlebrush Polymers by Broadband Dielectric Spectroscopy. Macromolecules 2020,* https://doi.org/10.1021/acs.macromol.0c01277)

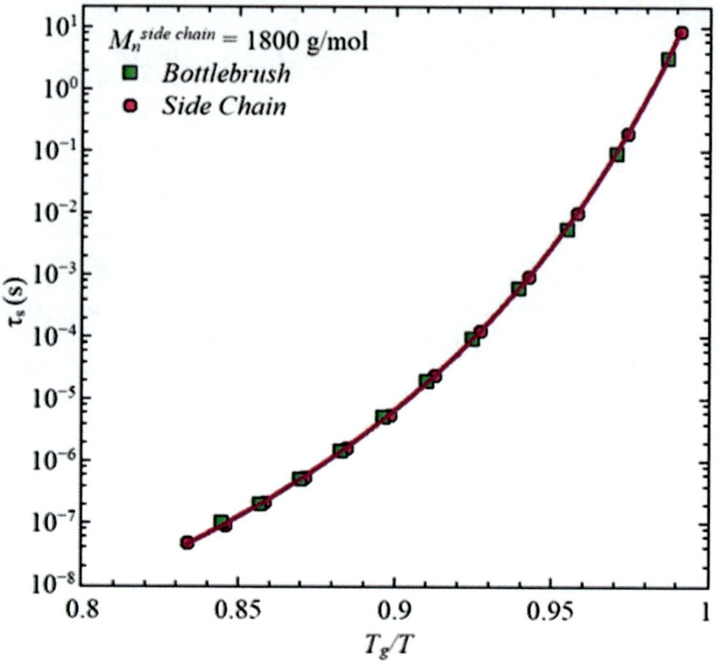

Fig. B.7 Segmental relaxation times, τ_s, vs. T_g/T for the PDMS-g-PDMS bottlebrush polymer and the single linear PDMS side chains of $M_n^{\text{side chain}} = 1800$ g/mol. Solid lines represent the description with the empirical VFT equation

Table B.7 Fit parameter by using the HN-function for describing α-relaxation of the linear PDMS side chains, $M_n^{\text{side chain}} = 1800$ g/mol, including the shape parameter β and γ, as well as the Havriliak–Negami relaxation time, τ_{HN}, and the real relaxation time, τ_s

	Temperature (°C)	β	γ	τ_{HN} (s)	τ_s (s)
α-relaxation	−127.5	0.74	0.48	21.75	8.91
	−125.0			0.49	0.20
	−122.5			0.03	0.01
	−120.0			$2.34 \cdot 10^{-3}$	$9.60 \cdot 10^{-4}$
	−117.5			$3.11 \cdot 10^{-4}$	$1.27 \cdot 10^{-4}$
	−115.0			$5.91 \cdot 10^{-5}$	$2.42 \cdot 10^{-5}$
	−112.5			$1.40 \cdot 10^{-5}$	$5.72 \cdot 10^{-5}$
	−110.0			$3.99 \cdot 10^{-6}$	$1.64 \cdot 10^{-6}$
	−107.5			$1.40 \cdot 10^{-6}$	$5.72 \cdot 10^{-7}$
	−105.0			$5.40 \cdot 10^{-7}$	$2.21 \cdot 10^{-7}$
	−102.5			$2.27 \cdot 10^{-7}$	$9.29 \cdot 10^{-8}$
	−100.0			$1.18 \cdot 10^{-8}$	$4.84 \cdot 10^{-8}$

All fit values have a standard deviation of around 1% and are omitted for the sake of clarity. (*Jakobi, B., Bichler, K. J., Sokolova, A., Schneider, G. J., Dynamics of PDMS-g-PDMS Bottlebrush Polymers by Broadband Dielectric Spectroscopy. Macromolecules 2020,* https://doi.org/10.1021/acs.macromol.0c01277)

PDMS-g-PDMS Bottlebrush with Long Side Chains, $M_n^{\text{side chain}} = 11,500$ *g/mol*

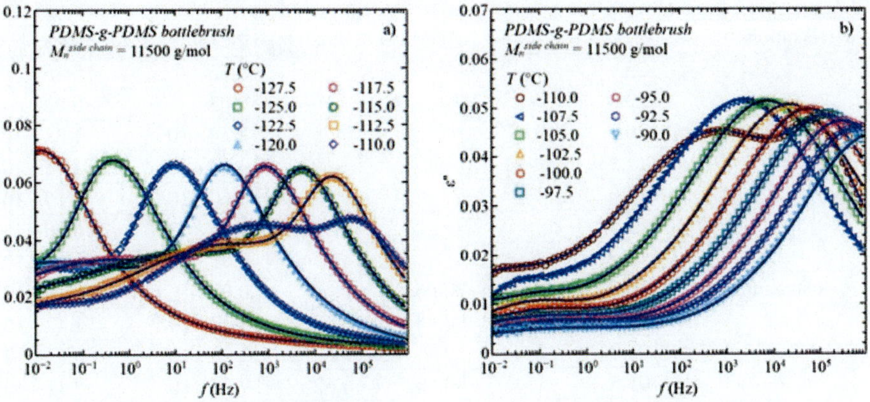

Fig. B.8 Imaginary part of the permittivity, ϵ'', vs. frequency, f, of the PDMS-g-PDMS bottlebrush polymer with $M_n^{\text{side chain}} = 11500$ g/mol. (**a**) The segmental relaxation, α-relaxation and (**b**) The segmental relaxation influenced by the cold crystallization, α_c-relaxation. Solid lines are the best description with the Havriliak–Negami function. (*Figure (**a**) taken from Jakobi, B., Bichler, K. J., Sokolova, A., Schneider, G. J., Dynamics of PDMS-g-PDMS Bottlebrush Polymers by Broadband Dielectric Spectroscopy. Macromolecules 2020,* https://doi.org/10.1021/acs.macromol.0c01277)

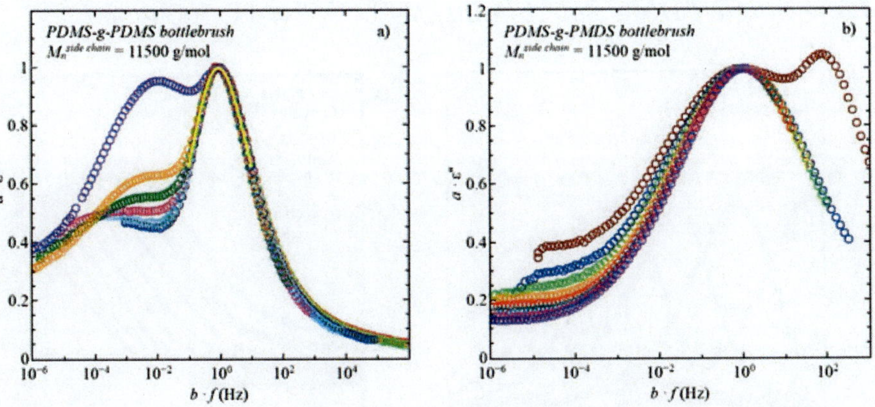

Fig. B.9 (**a**) Normalized imaginary part of the permittivity, $a \cdot \epsilon''$, vs. normalized frequency, $b \cdot f$, of the PDMS-g-PDMS bottlebrush polymer with $M_n^{\text{side chain}} = 11,500$ g/mol. (**a**) The segmental relaxation, α-relaxation, with the solid line as the best description with the Havriliak–Negami function and (**b**) The segmental relaxation influenced by the cold crystallization, α_c-relaxation. (*Figure (**a**) taken from Jakobi, B., Bichler, K. J., Sokolova, A., Schneider, G. J., Dynamics of PDMS-g-PDMS Bottlebrush Polymers by Broadband Dielectric Spectroscopy. Macromolecules 2020,* https://doi.org/10.1021/acs.macromol.0c01277)

Table B.8 Fit parameter for the HN-function for describing α-and α_c-relaxation of the PDMS-g-PDMS bottlebrush with side chains of $M_n^{\text{side chain}} = 11,500$ g/mol, including the shape parameter β and γ, as well as the Havriliak–Negami relaxation time, τ_{HN}, and the real relaxation time, τ_s

	Temperature (°C)	β	γ	τ_{HN} (s)	τ_s (s)
α-relaxation	−127.5	0.60	0.68	24.45	13.35
	−125			0.66	0.36
	−122.5			0.03	0.02
	−120.0			$2.54 \cdot 10^{-3}$	$1.38 \cdot 10^{-3}$
	−117.5			$3.19 \cdot 10^{-4}$	$1.74 \cdot 10^{-4}$
	−115.0			$5.60 \cdot 10^{-5}$	$3.10 \cdot 10^{-5}$
	−112.5			$1.26 \cdot 10^{-5}$	$6.86 \cdot 10^{-6}$
	−110.0			$4.82 \cdot 10^{-6}$	$2.63 \cdot 10^{-6}$
α_c-relaxation	−110.0	0.46	0.61	0.00019	$6.90 \cdot 10^{-5}$
	−107.5	0.48	0.58	$4.17 \cdot 10^{-5}$	$1.42 \cdot 10^{-5}$
	−105.0	0.49	0.53	$1.98 \cdot 10^{-5}$	$5.85 \cdot 10^{-6}$
	−102.5	0.50	0.52	$7.29 \cdot 10^{-6}$	$2.11 \cdot 10^{-6}$
	−100.0	0.51	0.50	$3.34 \cdot 10^{-6}$	$8.99 \cdot 10^{-7}$
	−97.5	0.50	0.45	$2.06 \cdot 10^{-6}$	$4.45 \cdot 10^{-7}$
	−95.0	0.50	0.42	$1.03 \cdot 10^{-6}$	$2.01 \cdot 10^{-7}$
	−92.5	0.49	0.38	$6.12 \cdot 10^{-7}$	$9.45 \cdot 10^{-7}$
	−90.0	0.49	0.35	$3.55 \cdot 10^{-7}$	$4.52 \cdot 10^{-7}$

All fit values have a standard deviation of around 1% and are omitted for the sake of clarity. (*Jakobi, B., Bichler, K. J., Sokolova, A., Schneider, G. J., Dynamics of PDMS-g-PDMS Bottlebrush Polymers by Broadband Dielectric Spectroscopy. Macromolecules 2020,* https://doi.org/10.1021/acs.macromol.0c01277)

Long Single PDMS Side Chains, $M_n^{\text{side chain}} = 11,500$ *g/mol*

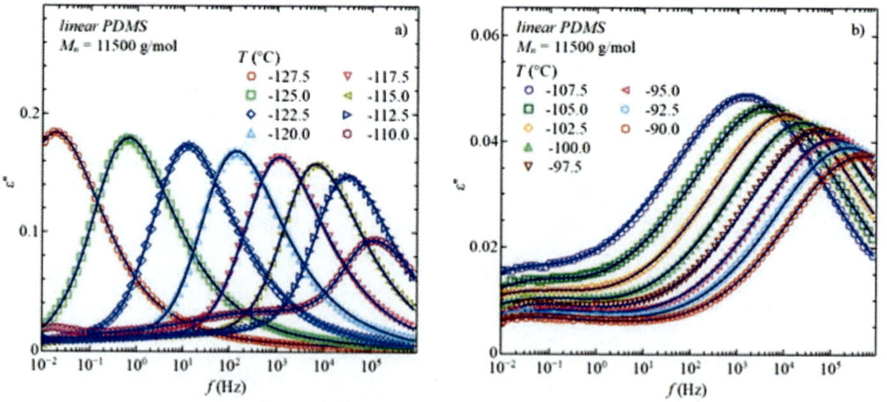

Fig. B.10 Imaginary part of the permittivity, ϵ'', vs. frequency, f, of the single linear PDMS side chains with $M_n = 11,500$ g/mol. (**a**) The segmental relaxation, α-relaxation and (**b**) The segmental relaxation influenced by the cold crystallization, α_c-relaxation. Solid lines are the best description with the Havriliak–Negami function. (*Figure (**a**) taken from Jakobi, B., Bichler, K. J., Sokolova, A., Schneider, G. J., Dynamics of PDMS-g-PDMS Bottlebrush Polymers by Broadband Dielectric Spectroscopy. Macromolecules 2020,* https://doi.org/10.1021/acs.macromol.0c01277)

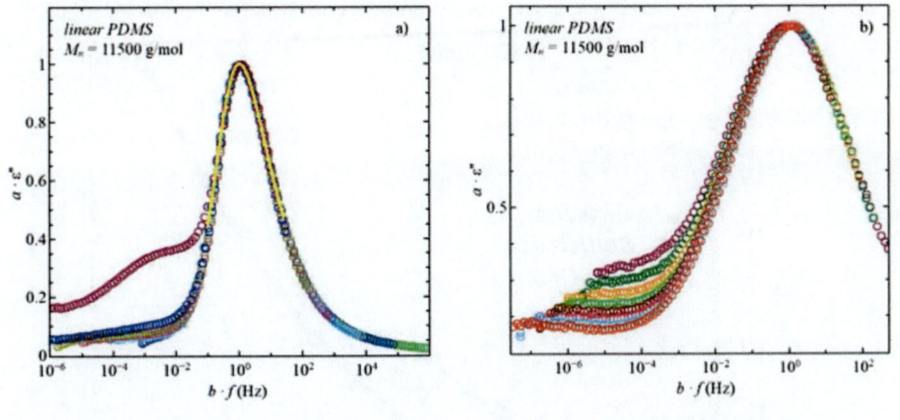

Fig. B.11 Normalized imaginary part of the permittivity, $a \cdot \epsilon''$, vs. normalized frequency, $b \cdot f$, of the single linear PDMS side chains with $M_n = 11,500$ g/mol. (**a**) The segmental relaxation, α-relaxation, with the solid line as the best description with the Havriliak–Negami function and (**b**) The segmental relaxation influenced by the cold crystallization, α_c-relaxation. (*Figure (**a**) taken from Jakobi, B., Bichler, K. J., Sokolova, A., Schneider, G. J., Dynamics of PDMS-g-PDMS Bottlebrush Polymers by Broadband Dielectric Spectroscopy. Macromolecules 2020,* https://doi.org/10.1021/acs.macromol.0c01277)

Table B.9 Fit parameter for the HN-function for describing α-and α_c-relaxation of the linear PDMS side chains of $M_n^{\text{side chain}} = 11,500$ g/mol, including the shape parameter β and γ, as well as the Havriliak–Negami relaxation time, τ_{HN}, and the real relaxation time, τ_s

	Temperature (°C)	β	γ	τ_{HN} (s)	τ_s (s)
α-relaxation	−127.5	0.80	0.46	22.73	9.68
	−125			0.57	0.24
	−122.5			0.03	0.01
	−120.0			$2.86 \cdot 10^{-3}$	$1.21 \cdot 10^{-3}$
	−117.5			$3.61 \cdot 10^{-4}$	$1.54 \cdot 10^{-4}$
	−115.0			$5.75 \cdot 10^{-5}$	$2.43 \cdot 10^{-5}$
	−112.5			$1.25 \cdot 10^{-6}$	$5.30 \cdot 10^{-6}$
	−110.0			$3.58 \cdot 10^{-6}$	$1.52 \cdot 10^{-6}$
α_c-relaxation	−107.5	0.47	0.57	$6.96 \cdot 10^{-5}$	$2.23 \cdot 10^{-5}$
	−105.0	0.49	0.54	$3.20 \cdot 10^{-5}$	$9.64 \cdot 10^{-6}$
	−102.5	0.50	0.51	$1.67 \cdot 10^{-5}$	$4.52 \cdot 10^{-6}$
	−100.0	0.50	0.47	$8.97 \cdot 10^{-6}$	$2.17 \cdot 10^{-6}$
	–97.5	0.50	0.43	$5.74 \cdot 10^{-6}$	$1.16 \cdot 10^{-7}$
	–95.0	0.50	0.41	$3.21 \cdot 10^{-6}$	$5.79 \cdot 10^{-7}$
	–92.5	0.50	0.33	$2.69 \cdot 10^{-6}$	$3.24 \cdot 10^{-7}$
	–90.0	0.49	0.28	$2.02 \cdot 10^{-6}$	$1.66 \cdot 10^{-7}$

All fit values have a standard deviation of around 1% and are omitted for the sake of clarity. (*Jakobi, B., Bichler, K. J., Sokolova, A., Schneider, G. J., Dynamics of PDMS-g-PDMS Bottlebrush Polymers by Broadband Dielectric Spectroscopy. Macromolecules 2020,* https://doi.org/10.1021/acs.macromol.0c01277)

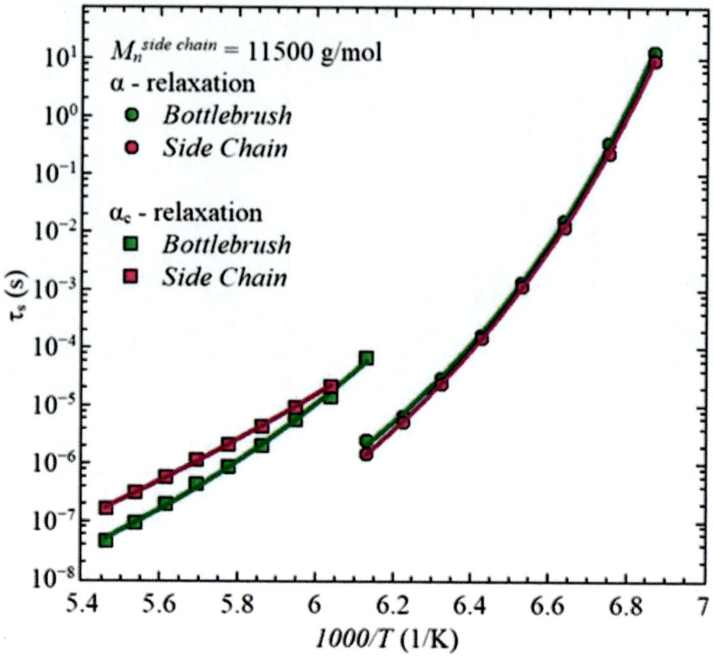

Fig. B.12 Segmental relaxation time, τ_s, vs. 1000/*T* of the PDMS-g-PDMS bottlebrush polymer and the single linear PDMS side chains with $M_n^{\text{side chain}} = 11,500$ g/mol. Solid lines are descriptions with the empirical VFT equation. (*Jakobi, B., Bichler, K. J., Sokolova, A., Schneider, G. J., Dynamics of PDMS-g-PDMS Bottlebrush Polymers by Broadband Dielectric Spectroscopy. Macromolecules 2020,* https://doi.org/10.1021/acs.macromol.0c01277)

Table B.10 Fit parameter of the VFT equation of the PDMS-g-PDMS bottlebrush with side chains of $M_n^{\text{side chain}} = 11,500$ g/mol and the respective side chains for the α_c-relaxation, including the limiting relaxation time at high temperatures, τ_∞, the fit parameter, *A*, and the ideal glass transition temperature, T_0

	τ_∞ (s)	*A* (K)	T_0 (K)
Bottlebrush	$(3 \pm 7) \cdot 10^{-14}$	863 ± 222	123 ± 6
Side chain	$(6 \pm 7) \cdot 10^{-16}$	1712 ± 182	95 ± 4

Fox–Loshaek Law

Describing the relationship between number average molecular weight, M_n, and the respective glass transition temperature, T_g, following expression was used.

$$\frac{1}{T_g} = \frac{1}{T_{g,\infty}} + \frac{K}{T_{g,\infty}^2 \cdot M_n} \tag{B.1}$$

Hereby, *K* as an empirical constant which is related to the free volume and $T_{g,\infty}$ describes the glass transition temperature for infinite molecular weights.

Quasi-Elastic Neutron Scattering on PDMS-g-PDMS Bottlebrush Polymers

PDMS-g-PDMS Bottlebrush with Short Side Chain, $M_n^{\text{side chain}} = 298$ g/mol

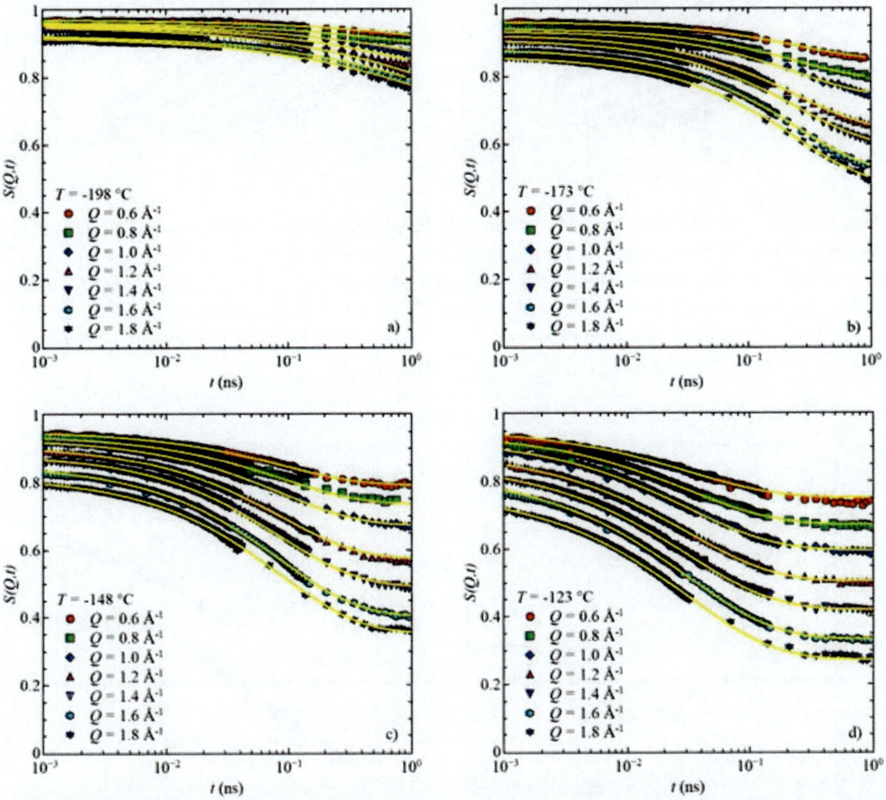

Fig. B.13 Intermediate scattering function, $S(Q, t)$, vs. time, t, of the PDMS-g-PDMS bottlebrush polymer with $M_n^{\text{side chain}} = 298$ g/mol. (**a**) -198 °C, (**b**) -173 °C, (**c**) -148 °C and (**d**) -123 °C. Solid lines are the best description with the model function for low temperatures. (*Bichler, K. J., Jakobi, B., García Sakai, V., Klapproth, A., Mole, R. A., Schneider, G. J., Short Time Dynamics of PDMS-g-PDMS Bottlebrush Polymer Melts Investigated by Quasi-Elastic Neutron Scattering. Macromolecules 2020,* https://doi.org/10.1021/acs.macromol.0c01846)

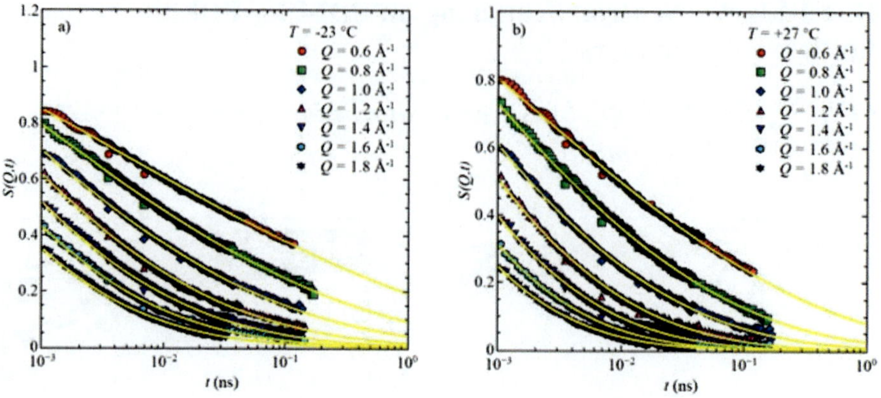

Fig. B.14 Intermediate scattering function, $S(Q, t)$, vs. time, t, of the PDMS-g-PDMS bottlebrush polymer with $M_n^{\text{side chain}} = 298$ g/mol. (**a**) −23 °C, and (**b**) +27 °C. Solid lines are the best description with the model function for high temperatures. (*Bichler, K. J., Jakobi, B., García Sakai, V., Klapproth, A., Mole, R. A., Schneider, G. J., Short Time Dynamics of PDMS-g-PDMS Bottlebrush Polymer Melts Investigated by Quasi-Elastic Neutron Scattering. Macromolecules 2020,* https://doi.org/10.1021/acs.macromol.0c01846)

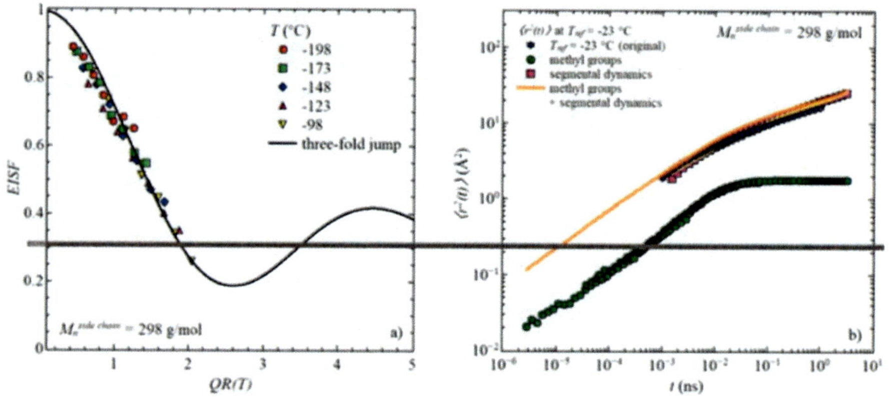

Fig. B.15 (**a**) Elastic incoherent structure factor, *EISF*, vs. $QR(T)$ for the PDMS-g-PDMS bottlebrush polymer with $M_n^{\text{side chain}} = 298$ g/mol. (**b**) Mean square displacement, $r^2(t)$, vs. time, t, for the PDMS-g-PDMS bottlebrush polymer with $M_n^{\text{side chain}} = 298$ g/mol. (*Figure (**b**) taken from Bichler, K. J., Jakobi, B., García Sakai, V., Klapproth, A., Mole, R. A., Schneider, G. J., Universality of the Time-Temperature Scaling Observed by Neutron Spectroscopy on Bottlebrush Polymers. Nano Letters 2021,* https://doi.org/10.1021/acs.nanolett.1c01379)

PDMS-g-PDMS Bottlebrush with Middle Side Chain, $M_n^{side\ chain} = 1800$ g/mol

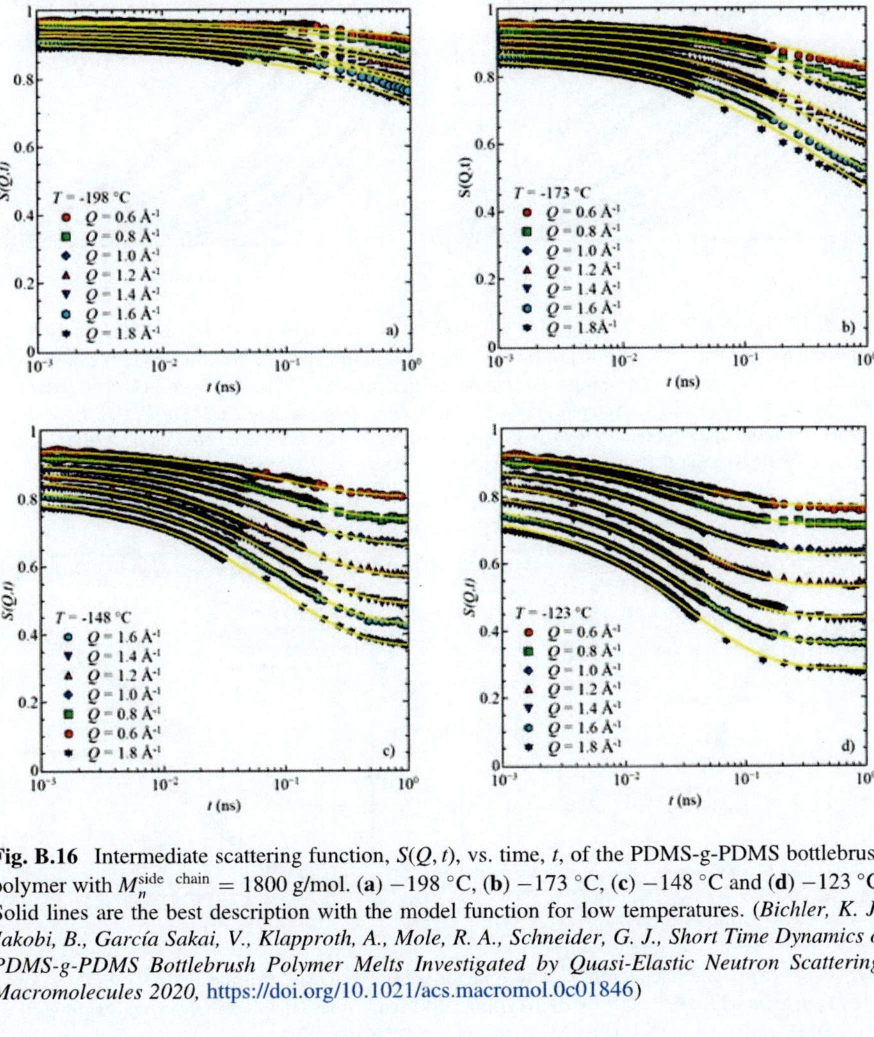

Fig. B.16 Intermediate scattering function, $S(Q, t)$, vs. time, t, of the PDMS-g-PDMS bottlebrush polymer with $M_n^{side\ chain} = 1800$ g/mol. (**a**) −198 °C, (**b**) −173 °C, (**c**) −148 °C and (**d**) −123 °C. Solid lines are the best description with the model function for low temperatures. (*Bichler, K. J., Jakobi, B., García Sakai, V., Klapproth, A., Mole, R. A., Schneider, G. J., Short Time Dynamics of PDMS-g-PDMS Bottlebrush Polymer Melts Investigated by Quasi-Elastic Neutron Scattering. Macromolecules 2020,* https://doi.org/10.1021/acs.macromol.0c01846)

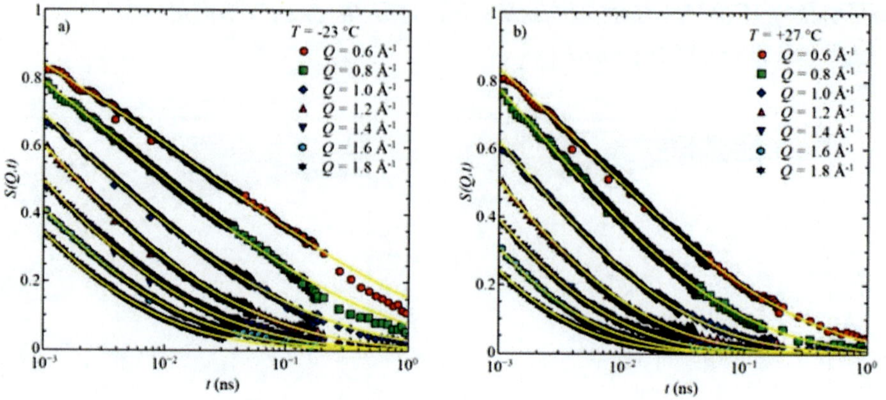

Fig. B.17 Intermediate scattering function, $S(Q, t)$, vs. time, t, of the PDMS-g-PDMS bottlebrush polymer with $M_n^{\text{side chain}} = 1800$ g/mol. (**a**) -23 °C, and (**b**) +27 °C. Solid lines are the description with the model function for high temperatures. (*Bichler, K. J., Jakobi, B., García Sakai, V., Klapproth, A., Mole, R. A., Schneider, G. J., Short Time Dynamics of PDMS-g-PDMS Bottlebrush Polymer Melts Investigated by Quasi-Elastic Neutron Scattering. Macromolecules 2020,* https://doi.org/10.1021/acs.macromol.0c01846)

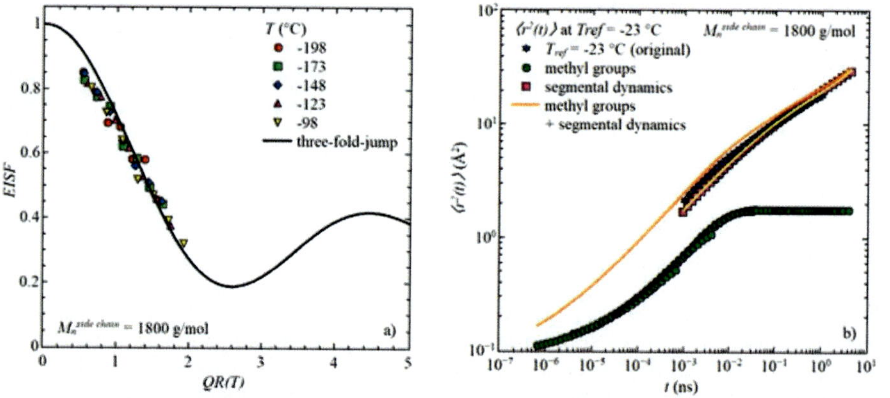

Fig. B.18 (**a**) Elastic incoherent structure factor, *EISF*, vs. $QR(T)$ for the PDMS-g-PDMS bottlebrush polymer with $M_n^{\text{side chain}} = 1800$ g/mol. (**b**) Mean square displacement, $r^2(t)$, vs. time, t, for the PDMS-g-PDMS bottlebrush polymer with $M_n^{\text{side chain}} = 1800$ g/mol. (*Figure (**b**) taken from Bichler, K. J., Jakobi, B., García Sakai, V., Klapproth, A., Mole, R. A., Schneider, G. J., Universality of the Time-Temperature Scaling Observed by Neutron Spectroscopy on Bottlebrush Polymers. Nano Letters 2021,* https://doi.org/10.1021/acs.nanolett.1c01379)

Appendix C: Publication Agreements and Permissions

Universality of the Time-Temperature Scaling Observed by Neutron Spectroscopy on Bottlebrush Polymers (https://doi.org/10.1021/acs.nanolett.1c01379)

ACS Chemistry for Life®

Dear Dr. Karin Bichler,

Thank you for contacting ACS Publications Help Desk.

Your permission requested is granted and there is no fee for this reuse. In your planned reuse, you must cite the ACS article as the source, https://pubs.acs.org/doi/abs/10.1021/acs.nanolett.1c01379, and further permissions related to the material excerpted should be directed to the ACS.

Should you have any questions or need further assistance, please feel free to contact me.

Sincerely,

Deleep Kumar
ACS Customer Services & Information

K.J. Bichler, *Morphology and Dynamics of Bottlebrush Polymers*, Springer Theses,
https://doi.org/10.1007/978-3-030-83379-4

Dynamical Comparison of Different Polymer Architectures: Bottlebrush Vs. Linear Polymer (https://doi.org/10.1021/acs.macromol.0c02104)

ACS
Chemistry for Life®

Dear Dr. Karin Bichler,

Thank you for contacting ACS Publications Help Desk.

Your permission requested is granted and there is no fee for this reuse. In your planned reuse, you must cite the ACS article as the source, https://pubs.acs.org/doi/10.1021/acs.macromol.0c02104, and further permissions related to the material excerpted should be directed to the ACS.

Should you have any questions or need further assistance, please feel free to contact me.

Sincerely,

Deleep Kumar
ACS Customer Services & Information

Short Time Dynamics of PDMS-g-PDMS Bottlebrush Polymer Melts Investigated by Quasi-Elastic Neutron Scattering (https://doi.org/10.1021/acs.macromol.0c01846)

ACS
Chemistry for Life®

Dear Dr. Karin Bichler,

Thank you for contacting ACS Publications Help Desk.

Your permission requested is granted and there is no fee for this reuse. In your planned reuse, you must cite the ACS article as the source, https://pubs.acs.org/doi/10.1021/acs.macromol.0c01846. Click or tap if you trust this link.">https://pubs.acs.org/doi/10.1021/acs.macromol.0c01846, and further permissions related to the material excerpted should be directed to the ACS.

Should you have any questions or need further assistance, please feel free to contact me.

Sincerely,

Deleep Kumar
ACS Customer Services & Information

Dynamics of PDMS-g-PDMS Bottlebrush Polymers by Broadband Dielectric Spectroscopy (https://doi.org/10.1021/acs.macromol.0c01277)

Copyright Clearance Center RightsLink®

Home Help Live Chat Sign in Create Account

ACS Publications Most Trusted. Most Cited. Most Read.

Dynamics of PDMS-g-PDMS Bottlebrush Polymers by Broadband Dielectric Spectroscopy

Author: Bruno Jakobi, Karin J. Bichler, Anna Sokolova, et al

Publication: Macromolecules

Publisher: American Chemical Society

Date: Oct 1, 2020

PERMISSION/LICENSE IS GRANTED FOR YOUR ORDER AT NO CHARGE

This type of permission/license, instead of the standard Terms & Conditions, is sent to you because no fee is being charged for your order. Please note the following:

- Permission is granted for your request in both print and electronic formats, and translations.
- If figures and/or tables were requested, they may be adapted or used in part.
- Please print this page for your records and send a copy of it to your publisher/graduate school.
- Appropriate credit for the requested material should be given as follows: "Reprinted (adapted) with permission from (COMPLETE REFERENCE CITATION). Copyright (YEAR) American Chemical Society." Insert appropriate information in place of the capitalized words.
- One-time permission is granted only for the use specified in your request. No additional uses are granted (such as derivative works or other editions). For any other uses, please submit a new request.

BACK CLOSE WINDOW

Structural Analysis of Ultrasoft PDMS-g-PDMS Shell-Only Particles (https://doi.org/10.1021/acs.macromol.9b01598)

Copyright Clearance Center RightsLink®

Home Help Email Support Sign in Create Account

ACS Publications
Most Trusted. Most Cited. Most Read.

Structural Analysis of Ultrasoft PDMS-g-PDMS Shell-Only Particles

Author: Karin J. Bichler, Bruno Jakobi, Stefan O. Huber, et al

Publication: Macromolecules

Publisher: American Chemical Society

Date: Jan 1, 2020

PERMISSION/LICENSE IS GRANTED FOR YOUR ORDER AT NO CHARGE

This type of permission/license, instead of the standard Terms & Conditions, is sent to you because no fee is being charged for your order. Please note the following:

- Permission is granted for your request in both print and electronic formats, and translations.
- If figures and/or tables were requested, they may be adapted or used in part.
- Please print this page for your records and send a copy of it to your publisher/graduate school.
- Appropriate credit for the requested material should be given as follows: "Reprinted (adapted) with permission from (COMPLETE REFERENCE CITATION). Copyright (YEAR) American Chemical Society." Insert appropriate information in place of the capitalized words.
- One-time permission is granted only for the use specified in your request. No additional uses are granted (such as derivative works or other editions). For any other uses, please submit a new request.

BACK CLOSE WINDOW